SAS® 9.3 SQL Procedure
User's Guide

SAS® Documentation

The correct bibliographic citation for this manual is as follows: SAS Institute Inc 2011. *SAS® 9.3 SQL Procedure User's Guide*. Cary, NC: SAS Institute Inc.

SAS® 9.3 SQL Procedure User's Guide

ISBN 978-1-60764-892-5

SAS Institute Inc., SAS Campus Drive, Cary, North Carolina 27513.

1st printing, July 2011

Contents

About This Book

Syntax Conventions for the SAS Language

Overview of Syntax Conventions for the SAS Language

SAS uses standard conventions in the documentation of syntax for SAS language elements. These conventions enable you to easily identify the components of SAS syntax. The conventions can be divided into these parts:

- syntax components
- style conventions
- special characters
- references to SAS libraries and external files

Syntax Components

The components of the syntax for most language elements include a keyword and arguments. For some language elements, only a keyword is necessary. For other language elements, the keyword is followed by an equal sign (=).

keyword

specifies the name of the SAS language element that you use when you write your program. Keyword is a literal that is usually the first word in the syntax. In a CALL routine, the first two words are keywords.

In the following examples of SAS syntax, the keywords are the first words in the syntax:

CHAR (*string, position*)

CALL RANBIN (*seed, n, p, x*);

ALTER (*alter-password*)

BEST *w*.

REMOVE *<data-set-name>*

In the following example, the first two words of the CALL routine are the keywords:

CALL RANBIN(*seed, n, p, x*)

The syntax of some SAS statements consists of a single keyword without arguments:

DO;
... SAS code ...

END;

Some system options require that one of two keyword values be specified:

DUPLEX | NODUPLEX

argument

specifies a numeric or character constant, variable, or expression. Arguments follow the keyword or an equal sign after the keyword. The arguments are used by SAS to process the language element. Arguments can be required or optional. In the syntax, optional arguments are enclosed between angle brackets.

In the following example, *string* and *position* follow the keyword CHAR. These arguments are required arguments for the CHAR function:

CHAR (*string, position*)

Each argument has a value. In the following example of SAS code, the argument *string* has a value of 'summer', and the argument *position* has a value of 4:`x=char('summer', 4);`

In the following example, *string* and *substring* are required arguments, while *modifiers* and *startpos* are optional.

FIND(*string, substring* <*,modifiers*> <*,startpos*>

Note: In most cases, example code in SAS documentation is written in lowercase with a monospace font. You can use uppercase, lowercase, or mixed case in the code that you write.

Style Conventions

The style conventions that are used in documenting SAS syntax include uppercase bold, uppercase, and italic:

UPPERCASE BOLD

identifies SAS keywords such as the names of functions or statements. In the following example, the keyword ERROR is written in uppercase bold:

ERROR<*message*>;

UPPERCASE

identifies arguments that are literals.

In the following example of the CMPMODEL= system option, the literals include BOTH, CATALOG, and XML:

CMPMODEL = BOTH | CATALOG | XML

italics

identifies arguments or values that you supply. Items in italics represent user-supplied values that are either one of the following:

- nonliteral arguments In the following example of the LINK statement, the argument *label* is a user-supplied value and is therefore written in italics:

  LINK *label*;

- nonliteral values that are assigned to an argument

  In the following example of the FORMAT statement, the argument DEFAULT is assigned the variable *default-format*:

  FORMAT = *variable-1* <, ..., *variable-nformat*><DEFAULT = *default-format*>;

Items in italics can also be the generic name for a list of arguments from which you can choose (for example, *attribute-list*). If more than one of an item in italics can be used, the items are expressed as *item-1, ..., item-n*.

Special Characters

The syntax of SAS language elements can contain the following special characters:

=

an equal sign identifies a value for a literal in some language elements such as system options.

In the following example of the MAPS system option, the equal sign sets the value of MAPS:

MAPS = *location-of-maps*

< >

angle brackets identify optional arguments. Any argument that is not enclosed in angle brackets is required.

In the following example of the CAT function, at least one item is required:

CAT (*item-1* <, ..., *item-n*>)

|

a vertical bar indicates that you can choose one value from a group of values. Values that are separated by the vertical bar are mutually exclusive.

In the following example of the CMPMODEL= system option, you can choose only one of the arguments:

CMPMODEL = BOTH | CATALOG | XML

...

an ellipsis indicates that the argument or group of arguments following the ellipsis can be repeated. If the ellipsis and the following argument are enclosed in angle brackets, then the argument is optional.

In the following example of the CAT function, the ellipsis indicates that you can have multiple optional items:

CAT (*item-1* <, ..., *item-n*>)

'*value*' or "*value*"

indicates that an argument enclosed in single or double quotation marks must have a value that is also enclosed in single or double quotation marks.

In the following example of the FOOTNOTE statement, the argument *text* is enclosed in quotation marks:

FOOTNOTE <*n*> <*ods-format-options* '*text*' | "*text*">;

;

a semicolon indicates the end of a statement or CALL routine.

In the following example each statement ends with a semicolon: `data namegame; length color name $8; color = 'black'; name = 'jack'; game = trim(color) || name; run;`

References to SAS Libraries and External Files

Many SAS statements and other language elements refer to SAS libraries and external files. You can choose whether to make the reference through a logical name (a libref or fileref) or use the physical filename enclosed in quotation marks. If you use a logical name, you usually have a choice of using a SAS statement (LIBNAME or FILENAME) or the operating environment's control language to make the association. Several methods of referring to SAS libraries and external files are available, and some of these methods depend on your operating environment.

In the examples that use external files, SAS documentation uses the italicized phrase *file-specification*. In the examples that use SAS libraries, SAS documentation uses the italicized phrase *SAS-library*. Note that *SAS-library* is enclosed in quotation marks:

```
infile file-specification obs = 100;
libname libref 'SAS-library';
```

What's New in the SAS 9.3 SQL Procedure

Overview

PROC SQL reference information from the *Base SAS Procedures Guide* and SAS SQL system options from the *SAS Language Reference: Dictionary* have been moved to this book, *SAS SQL Procedure User's Guide*. This enables our customers to access PROC SQL information in one location. The following are new features and enhancements:

- ability to optimize the PUT function
- ability to reuse the LIBNAME statement database connection
- additional PROC SQL statement options
- additional macro variable specifications for the INTO clause
- additional dictionary table
- additional system macro variable
- updated output examples

Ability to Optimize the PUT Function

The following reduce PUT options and system options have been modified to optimize the PUT function:

- REDUCEPUTOBS=
- REDUCEPUTVALUES=
- SQLREDUCEPUTOBS=

- SQLREDUCEPUTVALUES=

Ability to Reuse the LIBNAME Statement Database Connection

The database connection that is established with the LIBNAME statement can be reused in the CONNECT statement. The keyword USING has been added to implement this feature.

Additional PROC SQL Statement Options

The following PROC SQL statement options have been added to help control execution and output of results:

- STOPONTRUNC
- WARNRECURS | NOWARNRECURS

Additional Macro Variable Specifications for the INTO Clause

The following macro variable specifications have been added to the syntax for the INTO clause of the SELECT statement:

- TRIMMED option

- unbounded *macro-variable* range

Additional Dictionary Table

The VIEW_SOURCES dictionary table view has been added.

Additional System Macro Variable

The SYS_SQLSETLIMIT macro variable has been added for use with PROC SQL to improve database processing.

Updated Output Examples

Where applicable, all of the LISTING output examples have been updated to show the new ODS HTML output. The new SAS 9.3 output defaults apply only to the SAS windowing environment under Microsoft Windows and UNIX. For more information, see Chapter 1, "New Output Defaults in SAS 9.3," in *SAS Output Delivery System: User's Guide*.

Recommended Reading

- *Base SAS Procedures Guide*
- *Cody's Data Cleaning Techniques Using SAS Software*
- *Combining and Modifying SAS Data Sets: Examples*
- *SAS/GRAPH: Reference*
- *SAS Language Reference: Concepts*
- *SAS Language Reference: Dictionary*
- *SAS Macro Language: Reference*

For a complete list of SAS publications, go to support.sas.com/bookstore. If you have questions about which titles you need, please contact a SAS Publishing Sales Representative:

SAS Publishing Sales
SAS Campus Drive
Cary, NC 27513-2414
Phone: 1-800-727-3228
Fax: 1-919-677-8166
E-mail: sasbook@sas.com
Web address: support.sas.com/bookstore

Part 1

Using the SQL Procedure

Chapter 1
Introduction to the SQL Procedure

What Is SQL?

Structured Query Language (SQL) is a standardized, widely used language that retrieves and updates data in relational tables and databases.

A **relation** is a mathematical concept that is similar to the mathematical concept of a set. Relations are represented physically as two-dimensional tables that are arranged in rows and columns. Relational theory was developed by E. F. Codd, an IBM researcher, and first implemented at IBM in a prototype called System R. This prototype evolved into commercial IBM products based on SQL. The Structured Query Language is now in the public domain and is part of many vendors' products.

What Is the SQL Procedure?

The SQL procedure is the Base SAS implementation of Structured Query Language. PROC SQL is part of Base SAS software, and you can use it with any SAS data set (table). Often, PROC SQL can be an alternative to other SAS procedures or the DATA step. You can use SAS language elements such as global statements, data set options, functions, informats, and formats with PROC SQL just as you can with other SAS procedures. PROC SQL enables you to perform the following tasks:

- generate reports
- generate summary statistics
- retrieve data from tables or views

- combine data from tables or views
- create tables, views, and indexes
- update the data values in PROC SQL tables
- update and retrieve data from database management system (DBMS) tables
- modify a PROC SQL table by adding, modifying, or dropping columns

PROC SQL can be used in an interactive SAS session or within batch programs, and it can include global statements, such as TITLE and OPTIONS.

Terminology

Tables

A PROC SQL table is the same as a SAS data file. It is a SAS file of type DATA. PROC SQL tables consist of rows and columns. The rows correspond to observations in SAS data files, and the columns correspond to variables. The following table lists equivalent terms that are used in SQL, SAS, and traditional data processing.

Table 1.1 *Comparing Equivalent Terms*

SQL Term	SAS Term	Data Processing Term
table	SAS data file	file
row	observation	record
column	variable	field

You can create and modify tables by using the SAS DATA step, or by using the PROC SQL statements that are described in Chapter 4, "Creating and Updating Tables and Views," on page 109. Other SAS procedures and the DATA step can read and update tables that are created with PROC SQL.

SAS data files can have a one-level name or a two-level name. Typically, the names of temporary SAS data files have only one level, and the data files are stored in the WORK library. PROC SQL assumes that SAS data files that are specified with a one-level name are to be read from or written to the WORK library, unless you specify a USER library. You can assign a USER library with a LIBNAME statement or with the SAS system option USER=. For more information about how to work with SAS data files and libraries, see "Temporary and Permanent SAS Data Sets" in Chapter 2 of *Base SAS Procedures Guide.*

DBMS tables are tables that were created with other software vendors' database management systems. PROC SQL can connect to, update, and modify DBMS tables, with some restrictions. For more information, see "Accessing a DBMS with SAS/ACCESS Software" on page 162.

Queries

Queries retrieve data from a table, view, or DBMS. A query returns a query result, which consists of rows and columns from a table. With PROC SQL, you use a SELECT statement and its subordinate clauses to form a query. Chapter 2, "Retrieving Data from a Single Table," on page 19 describes how to build a query.

Views

PROC SQL views do not actually contain data as tables do. Rather, a PROC SQL view contains a stored SELECT statement or query. The query executes when you use the view in a SAS procedure or DATA step. When a view executes, it displays data that is derived from existing tables, from other views, or from SAS/ACCESS views. Other SAS procedures and the DATA step can use a PROC SQL view as they would any SAS data file. For more information about views, see Chapter 4, "Creating and Updating Tables and Views," on page 109.

Note: When you process PROC SQL views between a client and a server, getting the correct results depends on the compatibility between the client and server architecture. For more information, see "Accessing a SAS View" in Chapter 17 of *SAS/CONNECT User's Guide*.

Null Values

According to the ANSI Standard for SQL, a missing value is called a null value. It is not the same as a blank or zero value. However, to be compatible with the rest of SAS, PROC SQL treats missing values the same as blanks or zero values, and considers all three to be null values. This important concept comes up in several places in this document.

Comparing PROC SQL with the SAS DATA Step

PROC SQL can perform some of the operations that are provided by the DATA step and the PRINT, SORT, and SUMMARY procedures. The following query displays the total population of all the large countries (countries with population greater than 1 million) on each continent.

```
proc sql;
   title 'Population of Large Countries Grouped by Continent';
   select Continent, sum(Population) as TotPop format=comma15.
      from sql.countries
      where Population gt 1000000
      group by Continent
      order by TotPop;
quit;
```

Output 1.1 *Sample SQL Output*

Population of Large Countries Grouped by Continent

Continent	TotPop
Oceania	3,422,548
Australia	18,255,944
Central America and Caribbean	65,283,910
South America	316,303,397
North America	384,801,818
Africa	706,611,183
Europe	811,680,062
Asia	3,379,469,458

Here is a SAS program that produces the same result.

```
title 'Large Countries Grouped by Continent';
proc summary data=sql.countries;
   where Population > 1000000;
   class Continent;
   var Population;
   output out=sumPop sum=TotPop;
run;

proc sort data=SumPop;
   by totPop;
run;

proc print data=SumPop noobs;
   var Continent TotPop;
   format TotPop comma15.;
   where _type_=1;
run;
```

Output 1.2 *Sample DATA Step Output*

Large Countries Grouped by Continent

Continent	TotPop
Oceania	3,422,548
Australia	18,255,944
Central America and Caribbean	65,283,910
South America	316,303,397
North America	384,801,818
Africa	706,611,183
Europe	811,680,062
Asia	3,379,469,458

This example shows that PROC SQL can achieve the same results as Base SAS software but often with fewer and shorter statements. The SELECT statement that is shown in this example performs summation, grouping, sorting, and row selection. It also displays the query's results without the PRINT procedure.

PROC SQL executes without using the RUN statement. After you invoke PROC SQL you can submit additional SQL procedure statements without submitting the PROC statement again. Use the QUIT statement to terminate the procedure.

Notes about the Example Tables

For all examples, the following global statement is in effect:

```
libname sql 'SAS-data-library';
```

The tables that are used in this document contain geographic and demographic data. The data is intended to be used for the PROC SQL code examples only; it is not necessarily up-to-date or accurate.

Note: You can find instructions for downloading these data sets at **http://ftp.sas.com/samples/A56936**. These data sets are valid for SAS 9 as well as previous versions of SAS.

The COUNTRIES table contains data that pertains to countries. The Area column contains a country's area in square miles. The UNDate column contains the year a country entered the United Nations, if applicable.

Output 1.3 *COUNTRIES (Partial Output)*

COUNTRIES

Name	Capital	Population	Area	Continent	UNDate
Afghanistan	Kabul	17070323	251825	Asia	1946
Albania	Tirane	3407400	11100	Europe	1955
Algeria	Algiers	28171132	919595	Africa	1962
Andorra	Andorra la Vell	64634	200	Europe	1993
Angola	Luanda	9901050	481300	Africa	1976
Antigua and Barbuda	St. John's	65644	171	Central America	1981
Argentina	Buenos Aires	34248705	1073518	South America	1945
Armenia	Yerevan	3556864	11500	Asia	1992
Australia	Canberra	18255944	2966200	Australia	1945
Austria	Vienna	8033746	32400	Europe	1955
Azerbaijan	Baku	7760064	33400	Asia	1992
Bahamas	Nassau	275703	5400	Central America	1973
Bahrain	Manama	591800	300	Asia	1971
Bangladesh	Dhaka	1.2639E8	57300	Asia	1974
Barbados	Bridgetown	258534	200	Central America	1966

The WORLDCITYCOORDS table contains latitude and longitude data for world cities. Cities in the Western hemisphere have negative longitude coordinates. Cities in the

Southern hemisphere have negative latitude coordinates. Coordinates are rounded to the nearest degree.

Output 1.4 *WORLDCITYCOORDS (Partial Output)*

WORLDCITCOORDS

City	Country	Latitude	Longitude
Kabul	Afghanistan	35	69
Algiers	Algeria	37	3
Buenos Aires	Argentina	-34	-59
Cordoba	Argentina	-31	-64
Tucuman	Argentina	-27	-65
Adelaide	Australia	-35	138
Alice Springs	Australia	-24	134
Brisbane	Australia	-27	153
Darwin	Australia	-12	131
Melbourne	Australia	-38	145
Perth	Australia	-32	116
Sydney	Australia	-34	151
Vienna	Austria	48	16
Nassau	Bahamas	26	-77
Chittagong	Bangladesh	22	92

The USCITYCOORDS table contains the coordinates for cities in the United States. Because all cities in this table are in the Western hemisphere, all of the longitude coordinates are negative. Coordinates are rounded to the nearest degree.

Output 1.5 *USCITYCOORDS (Partial Output)*

USCITYCOORDS

City	State	Latitude	Longitude
Albany	NY	43	-74
Albuquerque	NM	36	-106
Amarillo	TX	35	-102
Anchorage	AK	61	-150
Annapolis	MD	39	-77
Atlanta	GA	34	-84
Augusta	ME	44	-70
Austin	TX	30	-98
Baker	OR	45	-118
Baltimore	MD	39	-76
Bangor	ME	45	-69
Baton Rouge	LA	31	-91
Birmingham	AL	33	-87
Bismarck	ND	47	-101
Boise	ID	43	-116

The UNITEDSTATES table contains data that is associated with the states. The Statehood column contains the date when the state was admitted into the Union.

Output 1.6 *UNITEDSTATES (Partial Output)*

UNITEDSTATES

Name	Capital	Population	Area	Continent	Statehood
Alabama	Montgomery	4447100	52423	North America	14DEC1819
Alaska	Juneau	626932	656400	North America	03JAN1959
Arizona	Phoenix	5130632	114000	North America	14FEB1912
Arkansas	Little Rock	2447996	53200	North America	15JUN1836
California	Sacramento	31518948	163700	North America	09SEP1850
Colorado	Denver	3601298	104100	North America	01AUG1876
Connecticut	Hartford	3405565	5500	North America	09JAN1788
Delaware	Dover	707232	2500	North America	07DEC1787
District of Colum	Washington	612907	100	North America	21FEB1871
Florida	Tallahassee	13814408	65800	North America	03MAR1845
Georgia	Atlanta	8186453	59400	North America	02JAN1788
Hawaii	Honolulu	1183198	10900	Oceania	21AUG1959
Idaho	Boise	1293953	83600	North America	03JUL1890
Illinois	Springfield	11813091	57900	North America	03DEC1818
Indiana	Indianapolis	5769553	36400	North America	11DEC1816

The POSTALCODES table contains postal code abbreviations.

Output 1.7 *POSTALCODES (Partial Output)*

POSTALCODES

Name	Code
Alabama	AL
Alaska	AK
American Samoa	AS
Arizona	AZ
Arkansas	AR
California	CA
Colorado	CO
Connecticut	CT
Delaware	DE
District Of Columbia	DC
Florida	FL
Georgia	GA
Guam	GU
Hawaii	HI
Idaho	ID

The WORLDTEMPS table contains average high and low temperatures from various international cities.

Output 1.8 *WORLDTEMPS (Partial Output)*

WORLDTEMPS

City	Country	AvgHigh	AvgLow
Algiers	Algeria	90	45
Amsterdam	Netherlands	70	33
Athens	Greece	89	41
Auckland	New Zealand	75	44
Bangkok	Thailand	95	69
Beijing	China	86	17
Belgrade	Yugoslavia	80	29
Berlin	Germany	75	25
Bogota	Colombia	69	43
Bombay	India	90	68
Bucharest	Romania	83	24
Budapest	Hungary	80	25
Buenos Aires	Argentina	87	48
Cairo	Egypt	95	48
Calcutta	India	97	56

The OILPROD table contains oil production statistics from oil-producing countries.

Output 1.9 *OILPROD (Partial Output)*

OILPROD

Country	BarrelsPerDay
Algeria	1,400,000
Canada	2,500,000
China	3,000,000
Egypt	900,000
Indonesia	1,500,000
Iran	4,000,000
Iraq	600,000
Kuwait	2,500,000
Libya	1,500,000
Mexico	3,400,000
Nigeria	2,000,000
Norway	3,500,000
Oman	900,000
Saudi Arabia	9,000,000
United States of America	8,000,000

The OILRSRVS table lists approximate oil reserves of oil-producing countries.

Output 1.10 *OILRSRVS (Partial Output)*

OILRSRVS

Country	Barrels
Algeria	9,200,000,000
Canada	7,000,000,000
China	25,000,000,000
Egypt	4,000,000,000
Gabon	1,000,000,000
Indonesia	5,000,000,000
Iran	90,000,000,000
Iraq	110,000,000,000
Kuwait	95,000,000,000
Libya	30,000,000,000
Mexico	50,000,000,000
Nigeria	16,000,000,000
Norway	11,000,000,000
Saudi Arabia	260,000,000,000
United Arab Emirates	100,000,000

The CONTINENTS table contains geographic data that relates to world continents.

Output 1.11 *CONTINENTS*

CONTINENTS

Name	Area	HighPoint	Height	LowPoint	Depth
Africa	11506000	Kilimanjaro	19340	Lake Assal	-512
Antarctica	5500000	Vinson Massif	16860		.
Asia	16988000	Everest	29028	Dead Sea	-1302
Australia	2968000	Kosciusko	7310	Lake Eyre	-52
Central America	.		.		.
Europe	3745000	El'brus	18510	Caspian Sea	-92
North America	9390000	McKinley	20320	Death Valley	-282
Oceania	.		.		.
South America	6795000	Aconcagua	22834	Valdes Peninsul	-131

The FEATURES table contains statistics that describe various types of geographical features, such as oceans, lakes, and mountains.

Output 1.12 *FEATURES (Partial Output)*

FEATURES

Name	Type	Location	Area	Height	Depth	Length
Aconcagua	Mountain	Argentina	.	22834	.	.
Amazon	River	South America	.	.	.	4000
Amur	River	Asia	.	.	.	2700
Andaman	Sea		218100	.	3667	.
Angel Falls	Waterfall	Venezuela	.	3212	.	.
Annapurna	Mountain	Nepal	.	26504	.	.
Aral Sea	Lake	Asia	25300	.	222	.
Ararat	Mountain	Turkey	.	16804	.	.
Arctic	Ocean		5105700	.	17880	.
Atlantic	Ocean		33420000	.	28374	.
Baffin	Island	Arctic	183810	.	.	.
Baltic	Sea		146500	.	180	.
Baykal	Lake	Russia	11780	.	5315	.
Bering	Sea		873000	.	4893	.
Black	Sea		196100	.	3906	.

Chapter 2

Retrieving Data from a Single Table

Overview of the SELECT Statement

How to Use the SELECT Statement

This chapter shows you how to perform the following tasks:

- retrieve data from a single table by using the SELECT statement
- validate the correctness of a SELECT statement by using the VALIDATE statement

With the SELECT statement, you can retrieve data from tables or data that is described by SAS data views.

Note: The examples in this chapter retrieve data from tables that are SAS data sets. However, you can use all of the operations that are described here with SAS data views.

The SELECT statement is the primary tool of PROC SQL. You use it to identify, retrieve, and manipulate columns of data from a table. You can also use several optional clauses within the SELECT statement to place restrictions on a query.

SELECT and FROM Clauses

The following simple SELECT statement is sufficient to produce a useful result:

```
select Name
   from sql.countries;
```

The SELECT statement must contain a SELECT clause and a FROM clause, both of which are required in a PROC SQL query. This SELECT statement contains the following:

- a SELECT clause that lists the Name column

- a FROM clause that lists the table in which the Name column resides

WHERE Clause

The WHERE clause enables you to restrict the data that you retrieve by specifying a condition that each row of the table must satisfy. PROC SQL output includes only those rows that satisfy the condition. The following SELECT statement contains a WHERE clause that restricts the query output to only those countries that have a population that is greater than 5,000,000 people:

```
select Name
   from sql.countries
   where Population gt 5000000;
```

ORDER BY Clause

The ORDER BY clause enables you to sort the output from a table by one or more columns. That is, you can put character values in either ascending or descending alphabetical order, and you can put numerical values in either ascending or descending numerical order. The default order is ascending. For example, you can modify the previous example to list the data by descending population:

```
select Name
   from sql.countries
   where Population gt 5000000
   order by Population desc;
```

GROUP BY Clause

The GROUP BY clause enables you to break query results into subsets of rows. When you use the GROUP BY clause, you use an aggregate function in the SELECT clause or a HAVING clause to instruct PROC SQL how to group the data. For details about aggregate functions, see "Summarizing Data" on page 56. PROC SQL calculates the aggregate function separately for each group. When you do not use an aggregate function, PROC SQL treats the GROUP BY clause as if it were an ORDER BY clause, and any aggregate functions are applied to the entire table.

The following query uses the SUM function to list the total population of each continent. The GROUP BY clause groups the countries by continent, and the ORDER BY clause puts the continents in alphabetical order:

```
select Continent, sum(Population)
   from sql.countries
   group by Continent
   order by Continent;
```

HAVING Clause

The HAVING clause works with the GROUP BY clause to restrict the groups in a query's results based on a given condition. PROC SQL applies the HAVING condition after grouping the data and applying aggregate functions. For example, the following query restricts the groups to include only the continents of Asia and Europe:

```
select Continent, sum(Population)
   from sql.countries
   group by Continent
   having Continent in ('Asia', 'Europe')
   order by Continent;
```

Ordering the SELECT Statement

When you construct a SELECT statement, you must specify the clauses in the following order:

1. SELECT
2. FROM
3. WHERE
4. GROUP BY
5. HAVING
6. ORDER BY

Note: Only the SELECT and FROM clauses are required.

The PROC SQL SELECT statement and its clauses are discussed in further detail in the following sections.

Selecting Columns in a Table

When you retrieve data from a table, you can select one or more columns by using variations of the basic SELECT statement.

Selecting All Columns in a Table

Use an asterisk in the SELECT clause to select all columns in a table. The following example selects all columns in the SQL.USCITYCOORDS table, which contains latitude and longitude values for U.S. cities:

```
libname sql 'SAS-library';

proc sql outobs=12;
   title 'U.S. Cities with Their States and Coordinates';
   select *
      from sql.uscitycoords;
```

Note: The OUTOBS= option limits the number of rows (observations) in the output. OUTOBS= is similar to the OBS= data set option. OUTOBS= is used throughout this document to limit the number of rows that are displayed in examples.

Note: In the tables used in these examples, latitude values that are south of the Equator are negative. Longitude values that are west of the Prime Meridian are also negative.

Output 2.1 *Selecting All Columns in a Table*

U.S. Cities with Their States and Coordinates

City	State	Latitude	Longitude
Albany	NY	43	-74
Albuquerque	NM	36	-106
Amarillo	TX	35	-102
Anchorage	AK	61	-150
Annapolis	MD	39	-77
Atlanta	GA	34	-84
Augusta	ME	44	-70
Austin	TX	30	-98
Baker	OR	45	-118
Baltimore	MD	39	-76
Bangor	ME	45	-69
Baton Rouge	LA	31	-91

Note: When you select all columns, PROC SQL displays the columns in the order in which they are stored in the table.

Selecting Specific Columns in a Table

To select a specific column in a table, list the name of the column in the SELECT clause. The following example selects only the City column in the SQL.USCITYCOORDS table:

```
libname sql 'SAS library';

proc sql outobs=12;
   title 'Names of U.S. Cities';
   select City
      from sql.uscitycoords;
```

Output 2.2 *Selecting One Column*

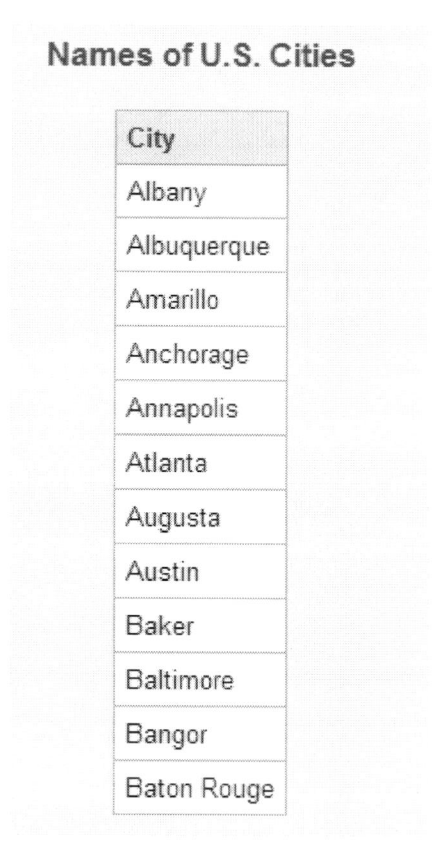

Names of U.S. Cities

City
Albany
Albuquerque
Amarillo
Anchorage
Annapolis
Atlanta
Augusta
Austin
Baker
Baltimore
Bangor
Baton Rouge

If you want to select more than one column, then you must separate the names of the columns with commas, as in this example, which selects the City and State columns in the SQL.USCITYCOORDS table:

```
libname sql 'SAS-library';

proc sql outobs=12;
   title 'U.S. Cities and Their States';
   select City, State
      from sql.uscitycoords;
```

Output 2.3 *Selecting Multiple Columns*

U.S. Cities and Their States

City	State
Albany	NY
Albuquerque	NM
Amarillo	TX
Anchorage	AK
Annapolis	MD
Atlanta	GA
Augusta	ME
Austin	TX
Baker	OR
Baltimore	MD
Bangor	ME
Baton Rouge	LA

Note: When you select specific columns, PROC SQL displays the columns in the order in which you specify them in the SELECT clause.

Eliminating Duplicate Rows from the Query Results

In some cases, you might want to find only the unique values in a column. For example, if you want to find the unique continents in which U.S. states are located, then you might begin by constructing the following query:

```
libname sql 'SAS-library';

proc sql outobs=12;
   title 'Continents of the United States';
   select Continent
      from sql.unitedstates;
```

Output 2.4 *Selecting a Column with Duplicate Values*

Continents of the United States

Continent
North America
North America
North America
North America
North America
North America
North America
North America
North America
North America
North America
Oceania

You can eliminate the duplicate rows from the results by using the DISTINCT keyword in the SELECT clause. Compare the previous example with the following query, which uses the DISTINCT keyword to produce a single row of output for each continent that is in the SQL.UNITEDSTATES table:

```
libname sql 'SAS-library';

proc sql;
   title 'Continents of the United States';
   select distinct Continent
      from sql.unitedstates;
```

Output 2.5 *Eliminating Duplicate Values*

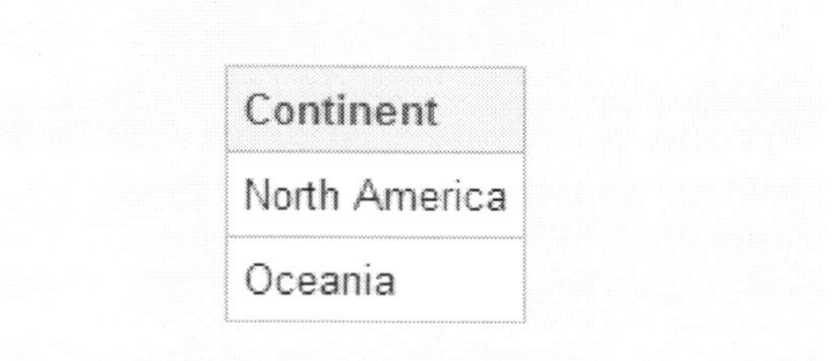

Continents of the United States

Continent
North America
Oceania

Note: When you specify all of a table's columns in a SELECT clause with the DISTINCT keyword, PROC SQL eliminates duplicate rows, or rows in which the values in all of the columns match, from the results.

Determining the Structure of a Table

To obtain a list of all of the columns in a table and their attributes, you can use the DESCRIBE TABLE statement. The following example generates a description of the SQL.UNITEDSTATES table. PROC SQL writes the description to the log.

```
libname sql 'SAS library';

proc sql;
   describe table sql.unitedstates;
```

Log 2.1 *Determining the Structure of a Table (Partial Log)*

```
NOTE: SQL table SQL.UNITEDSTATES was created like:

create table SQL.UNITEDSTATES( bufsize=12288 )
  (
   Name char(35) format=$35. informat=$35. label='Name',
   Capital char(35) format=$35. informat=$35. label='Capital',
   Population num format=BEST8. informat=BEST8. label='Population',
   Area num format=BEST8. informat=BEST8.,
   Continent char(35) format=$35. informat=$35. label='Continent',
   Statehood num
  );
```

Creating New Columns

In addition to selecting columns that are stored in a table, you can create new columns that exist for the duration of the query. These columns can contain text or calculations. PROC SQL writes the columns that you create as if they were columns from the table.

Adding Text to Output

You can add text to the output by including a string expression, or literal expression, in a query. The following query includes two strings as additional columns in the output:

```
libname sql 'SAS-library';

proc sql outobs=12;
   title 'U.S. Postal Codes';
   select 'Postal code for', Name, 'is', Code
      from sql.postalcodes;
```

Output 2.6 *Adding Text to Output*

U.S. Postal Codes

	Name		Code
Postal code for	Alabama	is	AL
Postal code for	Alaska	is	AK
Postal code for	American Samoa	is	AS
Postal code for	Arizona	is	AZ
Postal code for	Arkansas	is	AR
Postal code for	California	is	CA
Postal code for	Colorado	is	CO
Postal code for	Connecticut	is	CT
Postal code for	Delaware	is	DE
Postal code for	District Of Columbia	is	DC
Postal code for	Florida	is	FL
Postal code for	Georgia	is	GA

To prevent the column headings Name and Code from printing, you can assign a label that starts with a special character to each of the columns. PROC SQL does not output the column name when a label is assigned, and it does not output labels that begin with special characters. For example, you could use the following query to suppress the column headings that PROC SQL displayed in the previous example:

```
libname sql 'SAS-library';

proc sql outobs=12;
   title 'U.S. Postal Codes';
   select 'Postal code for', Name label='#', 'is', Code label='#'
      from sql.postalcodes;
```

Output 2.7 *Suppressing Column Headings in Output*

U.S. Postal Codes

Postal code for	Alabama	is	AL
Postal code for	Alaska	is	AK
Postal code for	American Samoa	is	AS
Postal code for	Arizona	is	AZ
Postal code for	Arkansas	is	AR
Postal code for	California	is	CA
Postal code for	Colorado	is	CO
Postal code for	Connecticut	is	CT
Postal code for	Delaware	is	DE
Postal code for	District Of Columbia	is	DC
Postal code for	Florida	is	FL
Postal code for	Georgia	is	GA

Calculating Values

You can perform calculations with values that you retrieve from numeric columns. The following example converts temperatures in the SQL.WORLDTEMPS table from Fahrenheit to Celsius:

```
libname sql 'SAS-library';

proc sql outobs=12;
   title 'Low Temperatures in Celsius';
   select City, (AvgLow - 32) * 5/9 format=4.1
      from sql.worldtemps;
```

Note: This example uses the FORMAT attribute to modify the format of the calculated output. For more information, see “Specifying Column Attributes” on page 36.

Output 2.8 *Calculating Values*

Low Temperatures in Celsius

City	
Algiers	7.2
Amsterdam	0.6
Athens	5.0
Auckland	6.7
Bangkok	20.6
Beijing	-8.3
Belgrade	-1.7
Berlin	-3.9
Bogota	6.1
Bombay	20.0
Bucharest	-4.4
Budapest	-3.9

Assigning a Column Alias

By specifying a column alias, you can assign a new name to any column within a PROC SQL query. The new name must follow the rules for SAS names. The name persists only for that query.

When you use an alias to name a column, you can use the alias to reference the column later in the query. PROC SQL uses the alias as the column heading in output. The following example assigns an alias of LowCelsius to the calculated column from the previous example:

```
libname sql 'SAS-library';

proc sql outobs=12;
   title 'Low Temperatures in Celsius';
   select City, (AvgLow - 32) * 5/9 as LowCelsius format=4.1
      from sql.worldtemps;
```

Output 2.9 *Assigning a Column Alias to a Calculated Column*

Low Temperatures in Celsius

City	LowCelsius
Algiers	7.2
Amsterdam	0.6
Athens	5.0
Auckland	6.7
Bangkok	20.6
Beijing	-8.3
Belgrade	-1.7
Berlin	-3.9
Bogota	6.1
Bombay	20.0
Bucharest	-4.4
Budapest	-3.9

Referring to a Calculated Column by Alias

When you use a column alias to refer to a calculated value, you must use the CALCULATED keyword with the alias to inform PROC SQL that the value is calculated within the query. The following example uses two calculated values, LowC and HighC, to calculate a third value, Range:

```
libname sql 'SAS-library';

proc sql outobs=12;
   title 'Range of High and Low Temperatures in Celsius';
      select City, (AvgHigh - 32) * 5/9 as HighC format=5.1,
                   (AvgLow - 32) * 5/9 as LowC format=5.1,
                   (calculated HighC - calculated LowC)
                    as Range format=4.1
   from sql.worldtemps;
```

Note: You can use an alias to refer to a calculated column in a SELECT clause, a WHERE clause, or ORDER BY clause.

Output 2.10 *Referring to a Calculated Column by Alias*

Range of High and Low Temperatures in Celsius

City	HighC	LowC	Range
Algiers	32.2	7.2	25.0
Amsterdam	21.1	0.6	20.6
Athens	31.7	5.0	26.7
Auckland	23.9	6.7	17.2
Bangkok	35.0	20.6	14.4
Beijing	30.0	-8.3	38.3
Belgrade	26.7	-1.7	28.3
Berlin	23.9	-3.9	27.8
Bogota	20.6	6.1	14.4
Bombay	32.2	20.0	12.2
Bucharest	28.3	-4.4	32.8
Budapest	26.7	-3.9	30.6

Note: Because this query sets a numeric format of 4.1 on the HighC, LowC, and Range columns, the values in those columns are rounded to the nearest tenth. As a result of the rounding, some of the values in the HighC and LowC columns do not reflect the range value output for the Range column. When you round numeric data values, this type of error sometimes occurs. If you want to avoid this problem, then you can specify additional decimal places in the format.

Assigning Values Conditionally

Using a Simple CASE Expression

CASE expressions enable you to interpret and change some or all of the data values in a column to make the data more useful or meaningful.

You can use conditional logic within a query by using a CASE expression to conditionally assign a value. You can use a CASE expression anywhere that you can use a column name.

The following table, which is used in the next example, describes the world climate zones (rounded to the nearest degree) that exist between Location 1 and Location 2:

Table 2.1 *World Climate Zones*

Climate zone	Location 1	Latitude at Location 1	Location 2	Latitude at Location 2
North Frigid	North Pole	90	Arctic Circle	67
North Temperate	Arctic Circle	67	Tropic of Cancer	23
Torrid	Tropic of Cancer	23	Tropic of Capricorn	-23
South Temperate	Tropic of Capricorn	-23	Antarctic Circle	-67
South Frigid	Antarctic Circle	-67	South Pole	-90

In this example, a CASE expression determines the climate zone for each city based on the value in the Latitude column in the SQL.WORLDCITYCOORDS table. The query also assigns an alias of ClimateZone to the value. You must close the CASE logic with the END keyword.

```
libname sql 'SAS-library';

proc sql outobs=12;
   title 'Climate Zones of World Cities';
   select City, Country, Latitude,
          case
             when Latitude gt 67 then 'North Frigid'
             when 67 ge Latitude ge 23 then 'North Temperate'
             when 23 gt Latitude gt -23 then 'Torrid'
             when -23 ge Latitude ge -67 then 'South Temperate'
             else 'South Frigid'
          end as ClimateZone
      from sql.worldcitycoords
      order by City;
```

Output 2.11 *Using a Simple CASE Expression*

Climate Zones of World Cities

City	Country	Latitude	ClimateZone
Abadan	Iran	30	North Temperate
Acapulco	Mexico	17	Torrid
Accra	Ghana	5	Torrid
Adana	Turkey	37	North Temperate
Addis Ababa	Ethiopia	9	Torrid
Adelaide	Australia	-35	South Temperate
Aden	Yemen	13	Torrid
Ahmenabad	India	22	Torrid
Algiers	Algeria	37	North Temperate
Alice Springs	Australia	-24	South Temperate
Amman	Jordan	32	North Temperate
Amsterdam	Netherlands	52	North Temperate

Using the CASE-OPERAND Form

You can also construct a CASE expression by using the CASE-OPERAND form, as in the following example. This example selects states and assigns them to a region based on the value of the Continent column:

```
libname sql 'SAS-library';

proc sql outobs=12;
   title 'Assigning Regions to Continents';
   select Name, Continent,
          case Continent
             when 'North America' then 'Continental U.S.'
             when 'Oceania' then 'Pacific Islands'
             else 'None'
          end as Region
      from sql.unitedstates;
```

Note: When you use the CASE-OPERAND form of the CASE expression, the conditions must all be equality tests. That is, they cannot use comparison operators or other types of operators, as are used in "Using a Simple CASE Expression" on page 32.

Output 2.12 *Using a CASE Expression in the CASE-OPERAND Form*

Assigning Regions to Continents

Name	Continent	Region
Alabama	North America	Continental U.S.
Alaska	North America	Continental U.S.
Arizona	North America	Continental U.S.
Arkansas	North America	Continental U.S.
California	North America	Continental U.S.
Colorado	North America	Continental U.S.
Connecticut	North America	Continental U.S.
Delaware	North America	Continental U.S.
District of Columbia	North America	Continental U.S.
Florida	North America	Continental U.S.
Georgia	North America	Continental U.S.
Hawaii	Oceania	Pacific Islands

Replacing Missing Values

The COALESCE function enables you to replace missing values in a column with a new value that you specify. For every row that the query processes, the COALESCE function checks each of its arguments until it finds a nonmissing value, and then returns that value. If all of the arguments are missing values, then the COALESCE function returns a missing value. For example, the following query replaces missing values in the LowPoint column in the SQL.CONTINENTS table with the words **`Not Available`**:

```
libname sql 'SAS-library';

proc sql;
   title 'Continental Low Points';
   select Name, coalesce(LowPoint, 'Not Available') as LowPoint
      from sql.continents;
```

Output 2.13 *Using the COALESCE Function to Replace Missing Values*

Continental Low Points

Name	LowPoint
Africa	Lake Assal
Antarctica	Not Available
Asia	Dead Sea
Australia	Lake Eyre
Central America and Caribbean	Not Available
Europe	Caspian Sea
North America	Death Valley
Oceania	Not Available
South America	Valdes Peninsula

The following CASE expression shows another way to perform the same replacement of missing values. However, the COALESCE function requires fewer lines of code to obtain the same results:

```
libname sql 'SAS-library';

proc sql;
   title 'Continental Low Points';
   select Name, case
                   when LowPoint is missing then 'Not Available'
                   else Lowpoint
                end as LowPoint
      from sql.continents;
```

Specifying Column Attributes

You can specify the following column attributes, which determine how SAS data is displayed:

- FORMAT=
- INFORMAT=
- LABEL=
- LENGTH=

If you do not specify these attributes, then PROC SQL uses attributes that are already saved in the table or, if no attributes are saved, then it uses the default attributes.

The following example assigns a label of `State` to the Name column and a format of COMMA10. to the Area column:

```
libname sql 'SAS-library';

proc sql outobs=12;
```

```
title 'Areas of U.S. States in Square Miles';
select Name label='State', Area format=comma10.
   from sql.unitedstates;
```

Note: Using the LABEL= keyword is optional. For example, the following two select clauses are the same:

```
select Name label='State', Area format=comma10.

select Name 'State', Area format=comma10.
```

Output 2.14 *Specifying Column Attributes*

Areas of U.S. States in Square Miles

State	Area
Alabama	52,423
Alaska	656,400
Arizona	114,000
Arkansas	53,200
California	163,700
Colorado	104,100
Connecticut	5,500
Delaware	2,500
District of Columbia	100
Florida	65,800
Georgia	59,400
Hawaii	10,900

Sorting Data

Overview of Sorting Data

You can sort query results with an ORDER BY clause by specifying any of the columns in the table, including columns that are not selected or columns that are calculated.

Unless an ORDER BY clause is included in the SELECT statement, then a particular order to the output rows, such as the order in which the rows are encountered in the queried table, cannot be guaranteed, even if an index is present. Without an ORDER BY clause, the order of the output rows is determined by the internal processing of PROC SQL, the default collating sequence of SAS, and your operating environment. Therefore, if you want your result table to appear in a particular order, then use the ORDER BY clause.

For more information and examples, see the “ORDER BY Clause” on page 301.

Sorting by Column

The following example selects countries and their populations from the SQL.COUNTRIES table and orders the results by population:

```
libname sql 'SAS-library';

proc sql outobs=12;
   title 'Country Populations';
   select Name, Population format=comma10.
      from sql.countries
      order by Population;
```

Note: When you use an ORDER BY clause, you change the order of the output but not the order of the rows that are stored in the table.

Note: The PROC SQL default sort order is ascending.

Output 2.15 *Sorting by Column*

Country Populations

Name	Population
Vatican City	1,010
Nauru	10,099
Tuvalu	10,099
Leeward Islands	12,119
Turks and Caicos Islands	12,119
Cayman Islands	23,228
San Marino	24,238
Liechtenstein	30,297
Gibraltar	30,297
Monaco	31,307
Saint Kitts and Nevis	41,406
Marshall Islands	54,535

Sorting by Multiple Columns

You can sort by more than one column by specifying the column names, separated by commas, in the ORDER BY clause. The following example sorts the SQL.COUNTRIES table by two columns, Continent and Name:

```
libname sql 'SAS-library';
```

```
proc sql outobs=12;
   title 'Countries, Sorted by Continent and Name';
   select Name, Continent
      from sql.countries
      order by Continent, Name;
```

Output 2.16 *Sorting by Multiple Columns*

Countries, Sorted by Continent and Name

Name	Continent
Bermuda	
Iceland	
Kalaallit Nunaat	
Algeria	Africa
Angola	Africa
Benin	Africa
Botswana	Africa
Burkina Faso	Africa
Burundi	Africa
Cameroon	Africa
Cape Verde	Africa
Central African Republic	Africa

Note: The results list countries without continents first because PROC SQL sorts missing values first in an ascending sort.

Specifying a Sort Order

To order the results, specify ASC for ascending or DESC for descending. You can specify a sort order for each column in the ORDER BY clause.

When you specify multiple columns in the ORDER BY clause, the first column determines the primary row order of the results. Subsequent columns determine the order of rows that have the same value for the primary sort. The following example sorts the SQL.FEATURES table by feature type and name:

```
libname sql 'SAS-library';

proc sql outobs=12;
   title 'World Topographical Features';
   select Name, Type
      from sql.features
      order by Type desc, Name;
```

Note: The ASC keyword is optional because the PROC SQL default sort order is ascending.

Output 2.17 *Specifying a Sort Order*

World Topographical Features

Name	Type
Angel Falls	Waterfall
Niagara Falls	Waterfall
Tugela Falls	Waterfall
Yosemite	Waterfall
Andaman	Sea
Baltic	Sea
Bering	Sea
Black	Sea
Caribbean	Sea
Gulf of Mexico	Sea
Hudson Bay	Sea
Mediterranean	Sea

Sorting by Calculated Column

You can sort by a calculated column by specifying its alias in the ORDER BY clause. The following example calculates population densities and then performs a sort on the calculated Density column:

```
libname sql 'SAS-library';

proc sql outobs=12;
   title 'World Population Densities per Square Mile';
   select Name, Population format=comma12., Area format=comma8.,
          Population/Area as Density format=comma10.
      from sql.countries
      order by Density desc;
```

Output 2.18 *Sorting by Calculated Column*

World Population Densities per Square Mile

Name	Population	Area	Density
Hong Kong	5,857,414	400	14,644
Singapore	2,887,301	200	14,437
Luxembourg	405,980	100	4,060
Malta	370,633	100	3,706
Maldives	254,495	100	2,545
Bangladesh	126,387,850	57,300	2,206
Bahrain	591,800	300	1,973
Taiwan	21,509,839	14,000	1,536
Channel Islands	146,436	100	1,464
Barbados	258,534	200	1,293
Korea, South	45,529,277	38,300	1,189
Mauritius	1,128,057	1,000	1,128

Sorting by Column Position

You can sort by any column within the SELECT clause by specifying its numerical position. By specifying a position instead of a name, you can sort by a calculated column that has no alias. The following example does not assign an alias to the calculated density column. Instead, the column position of 4 in the ORDER BY clause refers to the position of the calculated column in the SELECT clause:

```
libname sql 'SAS-library';

proc sql outobs=12;
   title 'World Population Densities per Square Mile';
   select Name, Population format=comma12., Area format=comma8.,
          Population/Area format=comma10. label='Density'
      from sql.countries
      order by 4 desc;
```

Note: PROC SQL uses a label, if one has been assigned, as a heading for a column that does not have an alias.

Output 2.19 *Sorting by Column Position*

World Population Densities per Square Mile

Name	Population	Area	Density
Hong Kong	5,857,414	400	14,644
Singapore	2,887,301	200	14,437
Luxembourg	405,980	100	4,060
Malta	370,633	100	3,706
Maldives	254,495	100	2,545
Bangladesh	126,387,850	57,300	2,206
Bahrain	591,800	300	1,973
Taiwan	21,509,839	14,000	1,536
Channel Islands	146,436	100	1,464
Barbados	258,534	200	1,293
Korea, South	45,529,277	38,300	1,189
Mauritius	1,128,057	1,000	1,128

Sorting by Columns That Are Not Selected

You can sort query results by columns that are not included in the query. For example, the following query returns all the rows in the SQL.COUNTRIES table and sorts them by population, even though the Population column is not included in the query:

```
libname sql 'SAS-library';

proc sql outobs=12;
   title 'Countries, Sorted by Population';
   select Name, Continent
      from sql.countries
      order by Population;
```

Output 2.20 *Sorting by Columns That Are Not Selected*

Countries, Sorted by Population

Name	Continent
Vatican City	Europe
Tuvalu	Oceania
Nauru	Oceania
Leeward Islands	Central America and Caribbean
Turks and Caicos Islands	Central America and Caribbean
Cayman Islands	Central America and Caribbean
San Marino	Europe
Liechtenstein	Europe
Gibraltar	Europe
Monaco	Europe
Saint Kitts and Nevis	Central America and Caribbean
Marshall Islands	Oceania

Specifying a Different Sorting Sequence

SORTSEQ= is a PROC SQL statement option that specifies the sorting sequence for PROC SQL to use when a query contains an ORDER BY clause. Use this option only if you want to use a sorting sequence other than your operating environment's default sorting sequence. Possible values include ASCII, EBCDIC, and some languages other than English. For example, in an operating environment that supports the EBCDIC sorting sequence, you could use the following option in the PROC SQL statement to set the sorting sequence to EBCDIC:

```
proc sql sortseq=ebcdic;
```

Note: SORTSEQ= affects only the ORDER BY clause. It does not override your operating environment's default comparison operations for the WHERE clause.

Operating Environment Information
See the SAS documentation for your operating environment for more information about the default and other sorting sequences for your operating environment.

Sorting Columns That Contain Missing Values

PROC SQL sorts nulls, or missing values, before character or numeric data. Therefore, when you specify ascending order, missing values appear first in the query results.

The following example sorts the rows in the CONTINENTS table by the LowPoint column:

```
libname sql 'SAS-library';

proc sql;
   title 'Continents, Sorted by Low Point';
   select Name, LowPoint
      from sql.continents
      order by LowPoint;
```

Because three continents have a missing value in the LowPoint column, those continents appear first in the output. Note that because the query does not specify a secondary sort, rows that have the same value in the LowPoint column, such as the first three rows of output, are not displayed in any particular order. In general, if you do not explicitly specify a sort order, then PROC SQL output is not guaranteed to be in any particular order.

Output 2.21 *Sorting Columns That Contain Missing Values*

Continents, Sorted by Low Point

Name	LowPoint
Central America and Caribbean	
Antarctica	
Oceania	
Europe	Caspian Sea
Asia	Dead Sea
North America	Death Valley
Africa	Lake Assal
Australia	Lake Eyre
South America	Valdes Peninsula

Retrieving Rows That Satisfy a Condition

The WHERE clause enables you to retrieve only rows from a table that satisfy a condition. WHERE clauses can contain any of the columns in a table, including columns that are not selected.

Using a Simple WHERE Clause

The following example uses a WHERE clause to find all countries that are in the continent of Europe and their populations:

```
libname sql 'SAS-library';

proc sql outobs=12;
   title 'Countries in Europe';
   select Name, Population format=comma10.
```

```
   from sql.countries
   where Continent = 'Europe';
```

Output 2.22 *Using a Simple WHERE Clause*

Countries in Europe

Name	Population
Albania	3,407,400
Andorra	64,634
Austria	8,033,746
Belarus	10,508,000
Belgium	10,162,614
Bosnia and Herzegovina	4,697,040
Bulgaria	8,887,111
Channel Islands	146,436
Croatia	4,744,505
Czech Republic	10,511,029
Denmark	5,239,356
England	49,293,170

Retrieving Rows Based on a Comparison

You can use comparison operators in a WHERE clause to select different subsets of data. The following table lists the comparison operators that you can use:

Table 2.2 *Comparison Operators*

Symbol	Mnemonic Equivalent	Definition	Example
=	EQ	equal to	`where Name = 'Asia';`
^= or ~= or ¬= or <>	NE	not equal to	`where Name ne 'Africa';`
>	GT	greater than	`where Area > 10000;`
<	LT	less than	`where Depth < 5000;`

Symbol	Mnemonic Equivalent	Definition	Example
>=	GE	greater than or equal to	**where Statehood >= '01jan1860'd;**
<=	LE	less than or equal to	**where Population <= 5000000;**

The following example subsets the SQL.UNITEDSTATES table by including only states with populations greater than 5,000,000 people:

```
libname sql 'SAS-library';

proc sql;
   title 'States with Populations over 5,000,000';
   select Name, Population format=comma10.
      from sql.unitedstates
      where Population gt 5000000
      order by Population desc;
```

Output 2.23 *Retrieving Rows Based on a Comparison*

States with Populations over 5,000,000

Name	Population
California	31,518,948
New York	18,377,334
Texas	18,209,994
Florida	13,814,408
Pennsylvania	12,167,566
Illinois	11,813,091
Ohio	11,200,790
Michigan	9,571,318
New Jersey	7,957,196
North Carolina	7,013,950
Georgia	6,985,572
Virginia	6,554,851
Massachusetts	6,071,816
Indiana	5,769,553
Washington	5,307,322
Missouri	5,285,610
Tennessee	5,149,273
Wisconsin	5,087,770
Maryland	5,014,048

Retrieving Rows That Satisfy Multiple Conditions

You can use logical, or Boolean, operators to construct a WHERE clause that contains two or more expressions. The following table lists the logical operators that you can use:

Table 2.3 *Logical (Boolean) Operators*

Symbol	Mnemonic Equivalent	Definition	Example
&	AND	specifies that both the previous and following conditions must be true	`Continent = 'Asia' and Population > 5000000`

Symbol	Mnemonic Equivalent	Definition	Example
! or | or ¦	OR	specifies that either the previous or the following condition must be true	`Population < 1000000 or Population > 5000000`
^ or ~ or ¬	NOT	specifies that the following condition must be false	`Continent not 'Africa'`

The following example uses two expressions to include only countries that are in Africa and that have a population greater than 20,000,000 people:

```
libname sql 'SAS-library';

proc sql;
   title 'Countries in Africa with Populations over 20,000,000';
   select Name, Population format=comma10.
      from sql.countries
      where Continent = 'Africa' and Population gt 20000000
      order by Population desc;
```

Output 2.24 *Retrieving Rows That Satisfy Multiple Conditions*

Countries in Africa with Populations over 20,000,000

Name	Population
Nigeria	99,062,003
Egypt	59,912,259
Ethiopia	59,291,170
South Africa	44,365,873
Congo, Democratic Republic of	43,106,529
Sudan	29,711,229
Morocco	28,841,705
Kenya	28,520,558
Tanzania	28,263,033
Algeria	28,171,132
Uganda	20,055,584

Note: You can use parentheses to improve the readability of WHERE clauses that contain multiple, or compound, expressions, such as the following:

```
where (Continent = 'Africa' and Population gt 2000000) or
      (Continent = 'Asia' and Population gt 1000000)
```

Using Other Conditional Operators

Overview of Using Other Conditional Operators

You can use many different conditional operators in a WHERE clause. The following table lists other operators that you can use:

Table 2.4 *Conditional Operators*

Operator	Definition	Example
ANY	specifies that at least one of a set of values obtained from a subquery must satisfy a given condition	`where Population > any (select Population from sql.countries)`
ALL	specifies that all of the values obtained from a subquery must satisfy a given condition	`where Population > all (select Population from sql.countries)`
BETWEEN-AND	tests for values within an inclusive range	`where Population between 1000000 and 5000000`
CONTAINS	tests for values that contain a specified string	`where Continent contains 'America';`
EXISTS	tests for the existence of a set of values obtained from a subquery	`where exists (select * from sql.oilprod);`
IN	tests for values that match one of a list of values	`where Name in ('Africa', 'Asia');`
IS NULL or IS MISSING	tests for missing values	`where Population is missing;`
LIKE	tests for values that match a specified pattern[1]	`where Continent like 'A %';`
=*	tests for values that sound like a specified value	`where Name =* 'Tiland';`

Note: All of these operators can be prefixed with the NOT operator to form a negative condition.

1 You can use a percent symbol (%) to match any number of characters. You can use an underscore (_) to match one arbitrary character.

Using the IN Operator

The IN operator enables you to include values within a list that you supply. The following example uses the IN operator to include only the mountains and waterfalls in the SQL.FEATURES table:

```
libname sql 'SAS-library';

proc sql outobs=12;
   title 'World Mountains and Waterfalls';
   select Name, Type, Height format=comma10.
      from sql.features
      where Type in ('Mountain', 'Waterfall')
      order by Height;
```

Output 2.25 *Using the IN Operator*

World Mountains and Waterfalls

Name	Type	Height
Niagara Falls	Waterfall	193
Yosemite	Waterfall	2,425
Tugela Falls	Waterfall	3,110
Angel Falls	Waterfall	3,212
Kosciusko	Mountain	7,310
Pico Duarte	Mountain	10,417
Cook	Mountain	12,349
Matterhorn	Mountain	14,690
Wilhelm	Mountain	14,793
Mont Blanc	Mountain	15,771
Ararat	Mountain	16,804
Vinson Massif	Mountain	16,864

Using the IS MISSING Operator

The IS MISSING operator enables you to identify rows that contain columns with missing values. The following example selects countries that are not located on a continent. That is, these countries have a missing value in the Continent column:

```
proc sql;
   title 'Countries with Missing Continents';
   select Name, Continent
      from sql.countries
      where Continent is missing;
```

Note: The IS NULL operator is the same as, and interchangeable with, the IS MISSING operator.

Output 2.26 *Using the IS MISSING Operator*

Countries with Missing Continents

Name	Continent
Bermuda	
Iceland	
Kalaallit Nunaat	

Using the BETWEEN-AND Operators

To select rows based on a range of values, you can use the BETWEEN-AND operators. This example selects countries that have latitudes within five degrees of the Equator:

```
proc sql outobs=12;
   title 'Equatorial Cities of the World';
   select City, Country, Latitude
      from sql.worldcitycoords
      where Latitude between -5 and 5;
```

Note: In the tables used in these examples, latitude values that are south of the Equator are negative. Longitude values that are west of the Prime Meridian are also negative.

Note: Because the BETWEEN-AND operators are inclusive, the values that you specify in the BETWEEN-AND expression are included in the results.

Output 2.27 *Using the BETWEEN-AND Operators*

Equatorial Cities of the World

City	Country	Latitude
Belem	Brazil	-1
Fortaleza	Brazil	-4
Bogota	Colombia	4
Cali	Colombia	3
Brazzaville	Congo	-4
Quito	Ecuador	0
Cayenne	French Guiana	5
Accra	Ghana	5
Medan	Indonesia	3
Palembang	Indonesia	-3
Nairobi	Kenya	-1
Kuala Lumpur	Malaysia	4

Using the LIKE Operator

The LIKE operator enables you to select rows based on pattern matching. For example, the following query returns all countries in the SQL.COUNTRIES table that begin with the letter *Z* and are any number of characters long, or end with the letter *a* and are five characters long:

```
libname sql 'SAS-library';

proc sql;
   title1 'Country Names that Begin with the Letter "Z"';
   title2 'or Are 5 Characters Long and End with the Letter "a"';
   select Name
      from sql.countries
      where Name like 'Z%' or Name like '____a';
```

Output 2.28 *Using the LIKE Operator*

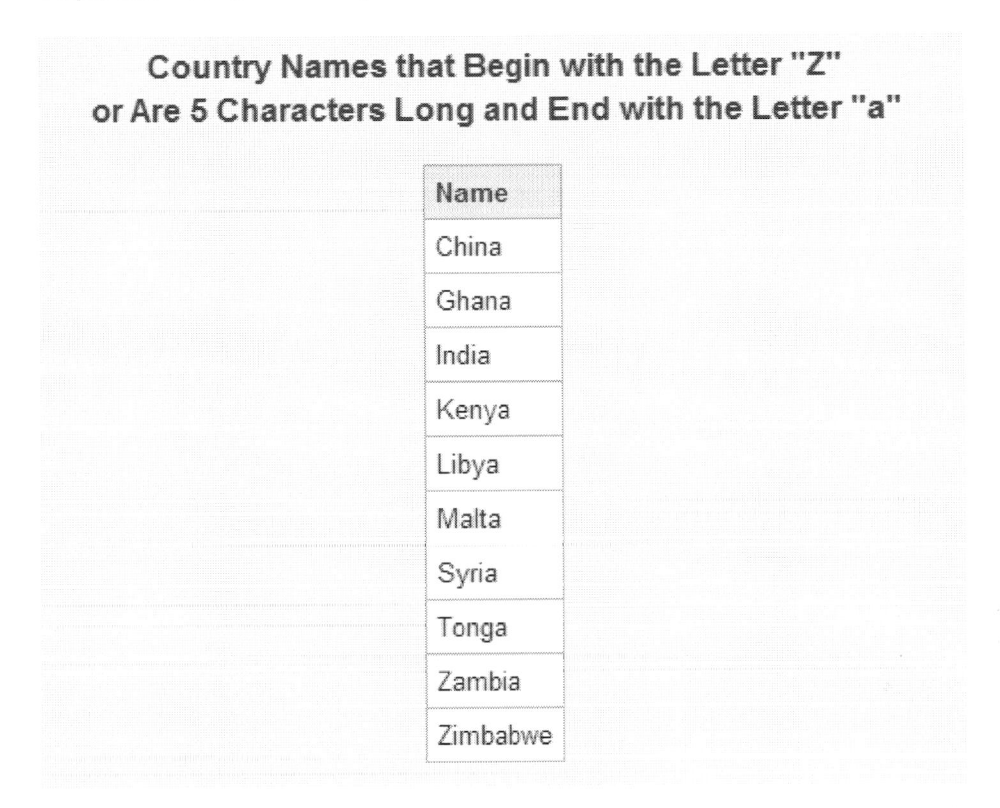
Country Names that Begin with the Letter "Z"
or Are 5 Characters Long and End with the Letter "a"

Name
China
Ghana
India
Kenya
Libya
Malta
Syria
Tonga
Zambia
Zimbabwe

The percent sign (%) and underscore (_) are wildcard characters. For more information about pattern matching with the LIKE comparison operator, see Chapter 7, "SQL Procedure," on page 209.

Using Truncated String Comparison Operators

Truncated string comparison operators are used to compare two strings. They differ from conventional comparison operators in that, before executing the comparison, PROC SQL truncates the longer string to be the same length as the shorter string. The truncation is performed internally; neither operand is permanently changed. The following table lists the truncated comparison operators:

Table 2.5 *Truncated String Comparison Operators*

Symbol	Definition	Example
EQT	equal to truncated strings	`where Name eqt 'Aust';`
GTT	greater than truncated strings	`where Name gtt 'Bah';`
LTT	less than truncated strings	`where Name ltt 'An';`
GET	greater than or equal to truncated strings	`where Country get 'United A';`
LET	less than or equal to truncated strings	`where Lastname let 'Smith';`

Symbol	Definition	Example
NET	not equal to truncated strings	**where Style net 'TWO';**

The following example returns a list of U.S. states that have **'New '** at the beginning of their names:

```
proc sql;
   title '"New" U.S. States';
   select Name
      from sql.unitedstates
      where Name eqt 'New ';
```

Output 2.29 *Using a Truncated String Comparison Operator*

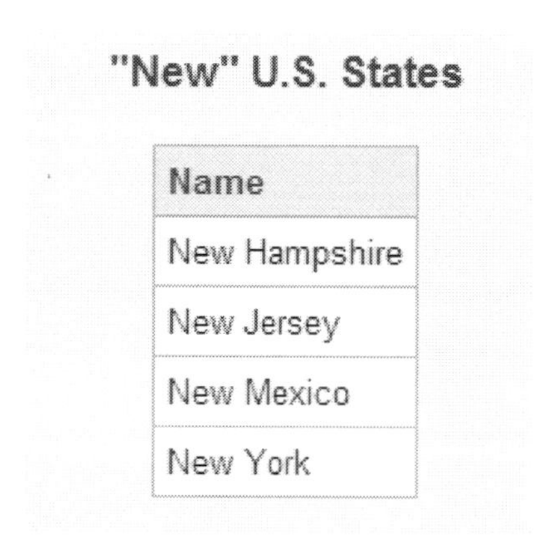

Name
New Hampshire
New Jersey
New Mexico
New York

Using a WHERE Clause with Missing Values

If a column that you specify in a WHERE clause contains missing values, then a query might provide unexpected results. For example, the following query returns all features from the SQL.FEATURES table that have a depth of less than 500 feet:

```
libname sql 'SAS-library';

/* incorrect output */

proc sql outobs=12;
   title 'World Features with a Depth of Less than 500 Feet';
   select Name, Depth
      from sql.features
      where Depth lt 500
      order by Depth;
```

Output 2.30 *Using a WHERE Clause with Missing Values (Incorrect Output)*

World Features with a Depth of Less than 500 Feet

Name	Depth
Kalahari	.
Nile	.
Citlaltepec	.
Lena	.
Mont Blanc	.
Borneo	.
Rub al Khali	.
Amur	.
Yosemite	.
Cook	.
Mackenzie-Peace	.
Mekong	.

However, because PROC SQL treats missing values as smaller than nonmissing values, features that have no depth listed are also included in the results. To avoid this problem, you could adjust the WHERE expression to check for missing values and exclude them from the query results, as follows:

```
libname sql 'SAS-library';

/* corrected output */

proc sql outobs=12;
   title 'World Features with a Depth of Less than 500 Feet';
   select Name, Depth
      from sql.features
      where Depth lt 500 and Depth is not missing
      order by Depth;
```

Output 2.31 *Using a WHERE Clause with Missing Values (Corrected Output)*

World Features with a Depth of Less than 500 Feet

Name	Depth
Baltic	180
Aral Sea	222
Victoria	264
Hudson Bay	305
North	308

Summarizing Data

Overview of Summarizing Data

You can use an *aggregate function* (or summary function) to produce a statistical summary of data in a table. The aggregate function instructs PROC SQL in how to combine data in one or more columns. If you specify one column as the argument to an aggregate function, then the values in that column are calculated. If you specify multiple arguments, then the arguments or columns that are listed are calculated.

Note: When more than one argument is used within an SQL aggregate function, the function is no longer considered to be an SQL aggregate or summary function. If there is a like-named Base SAS function, then PROC SQL executes the Base SAS function and the results that are returned are based on the values for the current row. If no like-named Base SAS function exists, then an error will occur. For example, if you use multiple arguments for the AVG function, an error will occur because there is no AVG function for Base SAS.

When you use an aggregate function, PROC SQL applies the function to the entire table, unless you use a GROUP BY clause. You can use aggregate functions in the SELECT or HAVING clauses.

Note: See "Grouping Data" on page 64 for information about producing summaries of individual groups of data within a table.

Using Aggregate Functions

The following table lists the aggregate functions that you can use:

Table 2.6 *Aggregate Functions*

Function	Definition
AVG, MEAN	mean or average of values
COUNT, FREQ, N	number of nonmissing values

Function	Definition
CSS	corrected sum of squares
CV	coefficient of variation (percent)
MAX	largest value
MIN	smallest value
NMISS	number of missing values
PRT	probability of a greater absolute value of Student's `t`
RANGE	range of values
STD	standard deviation
STDERR	standard error of the mean
SUM	sum of values
SUMWGT	sum of the WEIGHT variable values[1]
T	Student's `t` value for testing the hypothesis that the population mean is zero
USS	uncorrected sum of squares
VAR	variance

Note: You can use most other SAS functions in PROC SQL, but they are not treated as aggregate functions.

Summarizing Data with a WHERE Clause

Overview of Summarizing Data with a WHERE Clause

You can use aggregate, or summary functions, by using a WHERE clause. For a complete list of the aggregate functions that you can use, see Table 2.6 on page 56.

Using the MEAN Function with a WHERE Clause

This example uses the MEAN function to find the annual mean temperature for each country in the SQL.WORLDTEMPS table. The WHERE clause returns countries with a mean temperature that is greater than 75 degrees.

```
libname sql 'SAS-library';

proc sql outobs=12;
   title 'Mean Temperatures for World Cities';
```

1 In the SQL procedure, each row has a weight of 1.

```
select City, Country, mean(AvgHigh, AvgLow)
       as MeanTemp
   from sql.worldtemps
   where calculated MeanTemp gt 75
   order by MeanTemp desc;
```

Note: You must use the CALCULATED keyword to reference the calculated column.

Output 2.32 *Using the MEAN Function with a WHERE Clause*

Mean Temperatures for World Cities

City	Country	MeanTemp
Lagos	Nigeria	82.5
Manila	Philippines	82
Bangkok	Thailand	82
Singapore	Singapore	81
Bombay	India	79
Kingston	Jamaica	78
San Juan	Puerto Rico	78
Calcutta	India	76.5
Havana	Cuba	76.5
Nassau	Bahamas	76.5

Displaying Sums

The following example uses the SUM function to return the total oil reserves for all countries in the SQL.OILRSRVS table:

```
libname sql 'SAS-library';

proc sql;
   title 'World Oil Reserves';
   select sum(Barrels) format=comma18. as TotalBarrels
      from sql.oilrsrvs;
```

Note: The SUM function produces a single row of output for the requested sum because no nonaggregate value appears in the SELECT clause.

Output 2.33 Displaying Sums

World Oil Reserves

TotalBarrels
878,300,000,000

Combining Data from Multiple Rows into a Single Row

In the previous example, PROC SQL combined information from multiple rows of data into a single row of output. Specifically, the world oil reserves for each country were combined to form a total for all countries. Combining, or rolling up, of rows occurs when the following conditions exist:

- The SELECT clause contains only columns that are specified within an aggregate function.
- The WHERE clause, if there is one, contains only columns that are specified in the SELECT clause.

Remerging Summary Statistics

The following example uses the MAX function to find the largest population in the SQL.COUNTRIES table and displays it in a column called MaxPopulation. Aggregate functions, such as the MAX function, can cause the same calculation to repeat for every row. This occurs whenever PROC SQL remerges data. Remerging occurs whenever any of the following conditions exist:

- The SELECT clause references a column that contains an aggregate function that is not listed in a GROUP BY clause.
- The SELECT clause references a column that contains an aggregate function and other column or columns that are not listed in the GROUP BY clause.
- One or more columns or column expressions that are listed in a HAVING clause are not included in a subquery or a GROUP BY clause.

In this example, PROC SQL writes the population of China, which is the largest population in the table:

```
libname sql 'SAS-library';

proc sql outobs=12;
   title 'Largest Country Populations';
   select Name, Population format=comma20.,
          max(Population) as MaxPopulation format=comma20.
     from sql.countries
     order by Population desc;
```

***Output 2.34** Using Aggregate Functions*

Largest Country Populations

Name	Population	MaxPopulation
China	1,202,215,077	1,202,215,077
India	929,009,120	1,202,215,077
United States	263,294,808	1,202,215,077
Indonesia	202,393,859	1,202,215,077
Brazil	160,310,357	1,202,215,077
Russia	151,089,979	1,202,215,077
Bangladesh	126,387,850	1,202,215,077
Japan	126,345,434	1,202,215,077
Pakistan	123,062,252	1,202,215,077
Nigeria	99,062,003	1,202,215,077
Mexico	93,114,708	1,202,215,077
Germany	81,890,690	1,202,215,077

In some cases, you might need to use an aggregate function so that you can use its results in another calculation. To do this, you need only to construct one query for PROC SQL to automatically perform both calculations. This type of operation also causes PROC SQL to remerge the data.

For example, if you want to find the percentage of the total world population that resides in each country, then you construct a single query that performs the following tasks:

- obtains the total world population by using the SUM function
- divides each country's population by the total world population

PROC SQL runs an internal query to find the sum and then runs another internal query to divide each country's population by the sum.

```
libname sql 'SAS-library';

proc sql outobs=12;
   title 'Percentage of World Population in Countries';
   select Name, Population format=comma14.,
          (Population / sum(Population) * 100) as Percentage
          format=comma8.2
      from sql.countries
      order by Percentage desc;
```

Note: When a query remerges data, PROC SQL displays a note in the log to indicate that data remerging has occurred.

Output 2.35 *Remerging Summary Statistics*

Percentage of World Population in Countries

Name	Population	Percentage
China	1,202,215,077	21.10
India	929,009,120	16.30
United States	263,294,808	4.62
Indonesia	202,393,859	3.55
Brazil	160,310,357	2.81
Russia	151,089,979	2.65
Bangladesh	126,387,850	2.22
Japan	126,345,434	2.22
Pakistan	123,062,252	2.16
Nigeria	99,062,003	1.74
Mexico	93,114,708	1.63
Germany	81,890,690	1.44

Using Aggregate Functions with Unique Values

Counting Unique Values

You can use DISTINCT with an aggregate function to cause the function to use only unique values from a column.

The following query returns the number of distinct, nonmissing continents in the SQL.COUNTRIES table:

```
libname sql 'SAS-library';

proc sql;
   title 'Number of Continents in the COUNTRIES Table';
   select count(distinct Continent) as Count
      from sql.countries;
```

Output 2.36 *Using DISTINCT with the COUNT Function*

Number of Continents in the COUNTRIES Table

Count
8

Note: You cannot use **`select count(distinct *)`** to count distinct rows in a table. This code generates an error because PROC SQL does not know which duplicate column values to eliminate.

Counting Nonmissing Values

Compare the previous example with the following query, which does not use the DISTINCT keyword. This query counts every nonmissing occurrence of a continent in the SQL.COUNTRIES table, including duplicate values:

```
libname sql 'SAS-library';

proc sql;
   title 'Countries for Which a Continent is Listed';
   select count(Continent) as Count
      from sql.countries;
```

Output 2.37 *Effect of Not Using DISTINCT with the COUNT Function*

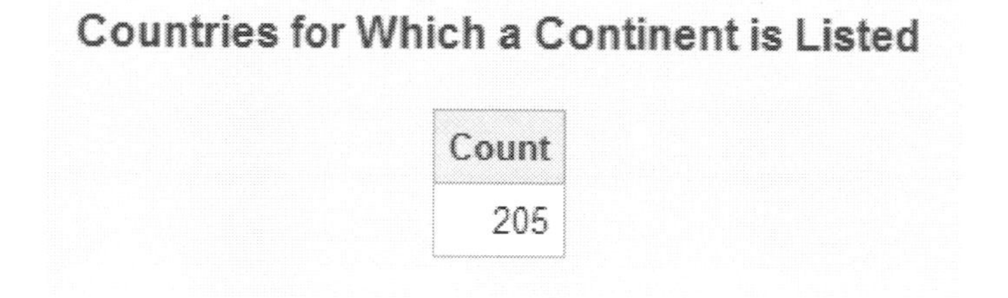
Countries for Which a Continent is Listed

Count
205

Counting All Rows

In the previous two examples, countries that have a missing value in the Continent column are ignored by the COUNT function. To obtain a count of all rows in the table, including countries that are not on a continent, you can use the following code in the SELECT clause:

```
proc sql;
   title 'Number of Countries in the SQL.COUNTRIES Table';
   select count(*) as Number
      from sql.countries;
```

Output 2.38 *Using the COUNT Function to Count All Rows in a Table*

Number of Countries in the SQL.COUNTRIES Table

Number
208

Summarizing Data with Missing Values

Overview of Summarizing Data with Missing Values

When you use an aggregate function with data that contains missing values, the results might not provide the information that you expect because many aggregate functions ignore missing values.

Finding Errors Caused by Missing Values

The AVG function returns the average of only the nonmissing values. The following query calculates the average length of three features in the SQL.FEATURES table: Angel Falls and the Amazon and Nile rivers:

```
libname sql 'SAS-library';

/* unexpected output */

proc sql;
   title 'Average Length of Angel Falls, Amazon and Nile Rivers';
   select Name, Length, avg(Length) as AvgLength
      from sql.features
      where Name in ('Angel Falls', 'Amazon', 'Nile');
```

Output 2.39 *Finding Errors Caused by Missing Values (Unexpected Output)*

Average Length of Angel Falls, Amazon and Nile Rivers

Name	Length	AvgLength
Amazon	4000	4072.5
Angel Falls	.	4072.5
Nile	4145	4072.5

Because no length is stored for Angel Falls, the average includes only the values for the Amazon and Nile rivers. Therefore, the average contains unexpected output results.

Compare the results from the previous example with the following query, which includes a COALESCE expression to handle missing values:

```
/* modified output */

proc sql;
   title 'Average Length of Angel Falls, Amazon and Nile Rivers';
   select Name, Length, coalesce(Length, 0) as NewLength,
                avg(calculated NewLength) as AvgLength
      from sql.features
      where Name in ('Angel Falls', 'Amazon', 'Nile');
```

Output 2.40 *Finding Errors Caused by Missing Values (Modified Output)*

Average Length of Angel Falls, Amazon and Nile Rivers

Name	Length	NewLength	AvgLength
Amazon	4000	4000	2715
Angel Falls	.	0	2715
Nile	4145	4145	2715

Grouping Data

The GROUP BY clause groups data by a specified column or columns. When you use a GROUP BY clause, you also use an aggregate function in the SELECT clause or in a HAVING clause to instruct PROC SQL in how to summarize the data for each group. PROC SQL calculates the aggregate function separately for each group.

Grouping by One Column

The following example sums the populations of all countries to find the total population of each continent:

```
libname sql 'SAS-library';

proc sql;
   title 'Total Populations of World Continents';
   select Continent, sum(Population) format=comma14. as TotalPopulation
      from sql.countries
      where Continent is not missing
      group by Continent;
```

Note: Countries for which a continent is not listed are excluded by the WHERE clause.

Output 2.41 *Grouping by One Column*

Total Populations of World Continents

Continent	TotalPopulation
Africa	710,529,592
Asia	3,381,858,879
Australia	18,255,944
Central America and Caribbean	66,815,930
Europe	813,481,724
North America	384,801,818
Oceania	5,342,368
South America	317,568,801

Grouping without Summarizing

When you use a GROUP BY clause without an aggregate function, PROC SQL treats the GROUP BY clause as if it were an ORDER BY clause and displays a message in the log that informs you that this has happened. The following example attempts to group high and low temperature information for each city in the SQL.WORLDTEMPS table by country:

```
libname sql 'SAS-library';

proc sql outobs=12;
   title 'High and Low Temperatures';
   select City, Country, AvgHigh, AvgLow
      from sql.worldtemps
      group by Country;
```

The output and log show that PROC SQL transforms the GROUP BY clause into an ORDER BY clause.

Output 2.42 *Grouping without Aggregate Functions*

High and Low Temperatures

City	Country	AvgHigh	AvgLow
Algiers	Algeria	90	45
Buenos Aires	Argentina	87	48
Sydney	Australia	79	44
Vienna	Austria	76	28
Nassau	Dahamas	88	65
Hamilton	Bermuda	85	59
Sao Paulo	Brazil	81	53
Rio de Janeiro	Brazil	85	64
Quebec	Canada	76	5
Montreal	Canada	77	8
Toronto	Canada	80	17
Beijing	China	86	17

Log 2.2 *Grouping without Aggregate Functions (Partial Log)*

```
WARNING: A GROUP BY clause has been transformed into an ORDER BY clause because

         neither the SELECT clause nor the optional HAVING clause of the
         associated table-expression referenced a summary function.
```

Grouping by Multiple Columns

To group by multiple columns, separate the column names with commas within the GROUP BY clause. You can use aggregate functions with any of the columns that you select. The following example groups by both Location and Type, producing total square miles for the deserts and lakes in each location in the SQL.FEATURES table:

```
libname sql 'SAS-library';

proc sql;
   title 'Total Square Miles of Deserts and Lakes';
   select Location, Type, sum(Area) as TotalArea format=comma16.
      from sql.features
      where type in ('Desert', 'Lake')
      group by Location, Type;
```

Output 2.43 *Grouping by Multiple Columns*

Total Square Miles of Deserts and Lakes

Location	Type	TotalArea
Africa	Desert	3,725,000
Africa	Lake	50,958
Asia	Lake	25,300
Australia	Desert	300,000
Canada	Lake	12,275
China	Desert	500,000
Europe - Asia	Lake	143,550
North America	Desert	140,000
North America	Lake	77,200
Russia	Lake	11,780
Saudi Arabia	Desert	250,000

Grouping and Sorting Data

You can order grouped results with an ORDER BY clause. The following example takes the previous example and adds an ORDER BY clause to change the order of the Location column from ascending order to descending order:

```
libname sql 'SAS-library';

proc sql;
   title 'Total Square Miles of Deserts and Lakes';
   select Location, Type, sum(Area) as TotalArea format=comma16.
      from sql.features
      where type in ('Desert', 'Lake')
      group by Location, Type
      order by Location desc;
```

Output 2.44 *Grouping with an ORDER BY Clause*

Total Square Miles of Deserts and Lakes

Location	Type	TotalArea
Saudi Arabia	Desert	250,000
Russia	Lake	11,780
North America	Lake	77,200
North America	Desert	140,000
Europe - Asia	Lake	143,550
China	Desert	500,000
Canada	Lake	12,275
Australia	Desert	300,000
Asia	Lake	25,300
Africa	Desert	3,725,000
Africa	Lake	50,958

Grouping with Missing Values

Finding Grouping Errors Caused by Missing Values

When a column contains missing values, PROC SQL treats the missing values as a single group. This can sometimes provide unexpected results.

In this example, because the SQL.COUNTRIES table contains some missing values in the Continent column, the missing values combine to form a single group that has the total area of the countries that have a missing value in the Continent column:

```
libname sql 'SAS-library';

/* unexpected output */

proc sql outobs=12;
   title 'Areas of World Continents';
   select Name format=$25.,
          Continent,
          sum(Area) format=comma12. as TotalArea
      from sql.countries
      group by Continent
      order by Continent, Name;
```

The output is incorrect because Bermuda, Iceland, and Kalaallit Nunaat are not actually part of the same continent. However, PROC SQL treats them that way because they all have a missing character value in the Continent column.

Output 2.45 *Finding Grouping Errors Caused by Missing Values (Unexpected Output)*

Areas of World Continents

Name	Continent	TotalArea
Bermuda		876,800
Iceland		876,800
Kalaallit Nunaat		876,800
Algeria	Africa	11,299,595
Angola	Africa	11,299,595
Benin	Africa	11,299,595
Botswana	Africa	11,299,595
Burkina Faso	Africa	11,299,595
Burundi	Africa	11,299,595
Cameroon	Africa	11,299,595
Cape Verde	Africa	11,299,595
Central African Republic	Africa	11,299,595

To correct the query from the previous example, you can write a WHERE clause to exclude the missing values from the results:

```
/* modified output */

proc sql outobs=12;
   title 'Areas of World Continents';
   select Name format=$25.,
          Continent,
          sum(Area) format=comma12. as TotalArea
      from sql.countries
      where Continent is not missing
      group by Continent
      order by Continent, Name;
```

Output 2.46 *Adjusting the Query to Avoid Errors Due to Missing Values (Modified Output)*

Areas of World Continents

Name	Continent	TotalArea
Algeria	Africa	11,299,595
Angola	Africa	11,299,595
Benin	Africa	11,299,595
Botswana	Africa	11,299,595
Burkina Faso	Africa	11,299,595
Burundi	Africa	11,299,595
Cameroon	Africa	11,299,595
Cape Verde	Africa	11,299,595
Central African Republic	Africa	11,299,595
Chad	Africa	11,299,595
Comoros	Africa	11,299,595
Congo	Africa	11,299,595

Note: Aggregate functions, such as the SUM function, can cause the same calculation to repeat for every row. This occurs whenever PROC SQL remerges data. See "Remerging Summary Statistics" on page 59 for more information about remerging.

Filtering Grouped Data

Overview of Filtering Grouped Data

You can use a HAVING clause with a GROUP BY clause to filter grouped data. The HAVING clause affects groups in a way that is similar to the way in which a WHERE clause affects individual rows. When you use a HAVING clause, PROC SQL displays only the groups that satisfy the HAVING expression.

Using a Simple HAVING Clause

The following example groups the features in the SQL.FEATURES table by type and then displays only the numbers of islands, oceans, and seas:

```
libname sql 'SAS-library';

proc sql;
   title 'Numbers of Islands, Oceans, and Seas';
   select Type, count(*) as Number
      from sql.features
      group by Type
```

```
having Type in ('Island', 'Ocean', 'Sea')
order by Type;
```

Output 2.47 *Using a Simple HAVING Clause*

Numbers of Islands, Oceans, and Seas

Type	Number
Island	6
Ocean	4
Sea	13

Choosing between HAVING and WHERE

The differences between the HAVING clause and the WHERE clause are shown in the following table. Because you use the HAVING clause when you work with groups of data, queries that contain a HAVING clause usually also contain the following:

- a GROUP BY clause
- an aggregate function

Note: When you use a HAVING clause without a GROUP BY clause, PROC SQL treats the HAVING clause as if it were a WHERE clause and provides a message in the log that informs you that this occurred.

Table 2.7 *Differences between the HAVING Clause and WHERE Clause*

HAVING clause attributes	WHERE clause attributes
is typically used to specify conditions for including or excluding groups of rows from a table.	is used to specify conditions for including or excluding individual rows from a table.
must follow the GROUP BY clause in a query, if used with a GROUP BY clause.	must precede the GROUP BY clause in a query, if used with a GROUP BY clause.
is affected by a GROUP BY clause, when there is no GROUP BY clause, the HAVING clause is treated like a WHERE clause.	is not affected by a GROUP BY clause.
is processed after the GROUP BY clause and any aggregate functions.	is processed before a GROUP BY clause, if there is one, and before any aggregate functions.

Using HAVING with Aggregate Functions

The following query returns the populations of all continents that have more than 15 countries:

```
libname sql 'SAS-library';

proc sql;
   title 'Total Populations of Continents with More than 15 Countries';
   select Continent,
          sum(Population) as TotalPopulation format=comma16.,
          count(*) as Count
      from sql.countries
      group by Continent
      having count(*) gt 15
      order by Continent;
```

The HAVING expression contains the COUNT function, which counts the number of rows within each group.

Output 2.48 *Using HAVING with the COUNT Function*

Total Populations of Continents with More than 15 Countries

Continent	TotalPopulation	Count
Africa	710,529,592	53
Asia	3,381,858,879	48
Central America and Caribbean	66,815,930	25
Europe	813,481,724	51

Validating a Query

The VALIDATE statement enables you to check the syntax of a query for correctness without submitting it to PROC SQL. PROC SQL displays a message in the log to indicate whether the syntax is correct.

```
libname sql 'SAS-library';

proc sql;
   validate
      select Name, Statehood
         from sql.unitedstates
         where Statehood lt '01Jan1800'd;
```

Log 2.3 *Validating a Query (Partial Log)*

```
3  proc sql;
4     validate
5        select Name, Statehood
6           from sql.unitedstates
7           where Statehood lt '01Jan1800'd;
NOTE: PROC SQL statement has valid syntax.
```

The following example shows an invalid query and the corresponding log message:

```
libname sql 'SAS-library';

proc sql;
   validate
      select Name, Statehood
      from sql.unitedstates
      where lt '01Jan1800'd;
```

Log 2.4 *Validating an Invalid Query (Partial Log)*

```
3  proc sql;
4     validate
5        select Name, Statehood
6        from sql.unitedstates
7        where lt '01Jan1800'd;
                  ------------
                  22
                  76
ERROR 22-322: Syntax error, expecting one of the following: !, !!, &, *, **,

              +, -, /, <, <=, <>, =, >, >=, ?, AND, CONTAINS, EQ, GE,
GROUP,
              GT, HAVING, LE, LIKE, LT, NE, OR, ORDER, ^=, |, ||, ~=.

ERROR 76-322: Syntax error, statement will be ignored.

NOTE: The SAS System stopped processing this step because of errors.
```

Chapter 3
Retrieving Data from Multiple Tables

Introduction

This chapter shows you how to perform the following tasks:

- select data from more than one table by joining the tables together
- use subqueries to select data from one table based on data values from another table
- combine the results of more than one query by using set operators

Note: Unless otherwise noted, the PROC SQL operations that are shown in this chapter apply to views as well as tables. For more information about views, see Chapter 4, "Creating and Updating Tables and Views," on page 109.

Selecting Data from More than One Table by Using Joins

Overview of Selecting Data from More than One Table by Using Joins

The data that you need for a report could be located in more than one table. In order to select the data from the tables, join the tables in a query. Joining tables enables you to select data from multiple tables as if the data were contained in one table. Joins do not alter the original tables.

The most basic type of join is simply two tables that are listed in the FROM clause of a SELECT statement. The following query joins the two tables that are shown in Output 3.1 on page 74 and creates Output 3.2 on page 75.

```
proc sql;
   title 'Table One and Table Two';
   select *
      from one, two;
```

```
proc sql;
  title 'Table One';
  select * from one;

  title 'Table Two';
  select * from two;

quit;
```

Output 3.1 *Table One, Table Two*

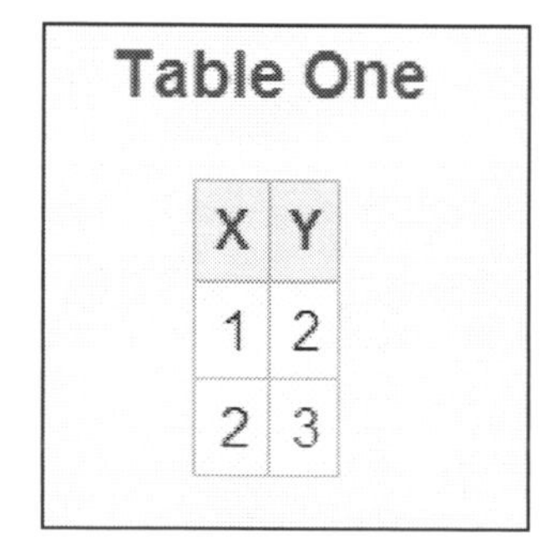

Table One

X	Y
1	2
2	3

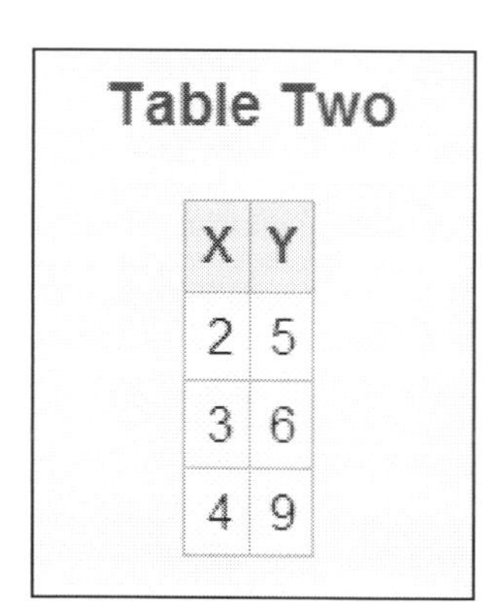

Table Two

X	Y
2	5
3	6
4	9

Output 3.2 *Cartesian Product of Table One and Table Two*

Table One and Table Two

X	Y	X	Y
1	2	2	5
1	2	3	6
1	2	4	9
2	3	2	5
2	3	3	6
2	3	4	9

Joining tables in this way returns the Cartesian product of the tables. Each row from the first table is combined with every row from the second table. When you run this query, the following message is written to the SAS log:

Log 3.1 *Cartesian Product Log Message*

```
NOTE: The execution of this query involves performing one or more Cartesian
      product joins that can not be optimized.
```

The Cartesian product of large tables can be huge. Typically, you want a subset of the Cartesian product. You specify the subset by declaring the join type.

There are two types of joins:

- Inner Joins return a result table for all the rows in a table that have one or more matching rows in the other table or tables that are listed in the FROM clause.
- Outer Joins are inner joins that are augmented with rows that did not match with any row from the other table in the join. There are three types of outer joins: left, right, and full.

Inner Joins

Overview of Inner Joins

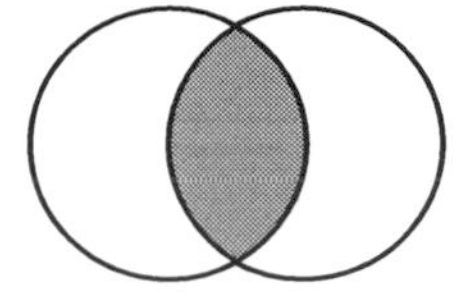

An inner join returns only the subset of rows from the first table that matches rows from the second table. You can specify the columns that you want to be compared for matching values in a WHERE clause.

The following code adds a WHERE clause to the previous query. The WHERE clause specifies that only rows whose values in column X of Table One match values in column

X of Table Two should appear in the output. Compare this query's output to Output 3.2 on page 75.

```
proc sql;
   title 'Table One and Table Two';
   select * from one, two
      where one.x=two.x;
```

Output 3.3 *Table One and Table Two Joined*

Table One and Table Two

X	Y	X	Z
2	3	2	5

The output contains only one row because only one value in column X matches from each table. In an inner join, only the matching rows are selected. Outer joins can return nonmatching rows; they are covered in "Outer Joins" on page 84.

Note that the column names in the WHERE clause are prefixed by their table names. This is known as qualifying the column names, and it is necessary when you specify columns that have the same name from more than one table. Qualifying the column name avoids creating an ambiguous column reference.

Using Table Aliases

A table alias is a temporary, alternate name for a table. You specify table aliases in the FROM clause. Table aliases are used in joins to qualify column names and can make a query easier to read by abbreviating table names.

The following example compares the oil production of countries to their oil reserves by joining the OILPROD and OILRSRVS tables on their Country columns. Because the Country columns are common to both tables, they are qualified with their table aliases. You could also qualify the columns by prefixing the column names with the table names.

Note: The AS keyword is optional.

```
libname sql 'SAS-library';

proc sql outobs=6;
   title 'Oil Production/Reserves of Countries';
   select * from sql.oilprod as p, sql.oilrsrvs as r
      where p.country = r.country;
```

Output 3.4 *Abbreviating Column Names by Using Table Aliases*

Oil Production/Reserves of Countries

Country	BarrelsPerDay	Country	Barrels
Algeria	1,400,000	Algeria	9,200,000,000
Canada	2,500,000	Canada	7,000,000,000
China	3,000,000	China	25,000,000,000
Egypt	900,000	Egypt	4,000,000,000
Indonesia	1,500,000	Indonesia	5,000,000,000
Iran	4,000,000	Iran	90,000,000,000

Note that each table's Country column is displayed. Typically, once you have determined that a join is functioning correctly, you include just one of the matching columns in the SELECT clause.

Specifying the Order of Join Output

You can order the output of joined tables by one or more columns from either table. The next example's output is ordered in descending order by the BarrelsPerDay column. It is not necessary to qualify BarrelsPerDay, because the column exists only in the OILPROD table.

```
libname sql 'SAS-library';

proc sql outobs=6;
   title 'Oil Production/Reserves of Countries';
   select p.country, barrelsperday 'Production', barrels 'Reserves'
      from sql.oilprod p, sql.oilrsrvs r
      where p.country = r.country
      order by barrelsperday desc;
```

Output 3.5 *Ordering the Output of Joined Tables*

Oil Production/Reserves of Countries

Country	Production	Reserves
Saudi Arabia	9,000,000	260,000,000,000
United States of America	8,000,000	30,000,000,000
Iran	4,000,000	90,000,000,000
Norway	3,500,000	11,000,000,000
Mexico	3,400,000	50,000,000,000
China	3,000,000	25,000,000,000

Creating Inner Joins Using INNER JOIN Keywords

The INNER JOIN keywords can be used to join tables. The ON clause replaces the WHERE clause for specifying columns to join. PROC SQL provides these keywords primarily for compatibility with the other joins (OUTER, RIGHT, and LEFT JOIN). Using INNER JOIN with an ON clause provides the same functionality as listing tables in the FROM clause and specifying join columns with a WHERE clause.

This code produces the same output as the previous code but uses the INNER JOIN construction.

```
proc sql ;
   select p.country, barrelsperday 'Production', barrels 'Reserves'
      from sql.oilprod p inner join sql.oilrsrvs r
           on p.country = r.country
   order by barrelsperday desc;
```

Joining Tables Using Comparison Operators

Tables can be joined by using comparison operators other than the equal sign (=) in the WHERE clause. For more information about comparison operators, see "Retrieving Rows Based on a Comparison" on page 45. In this example, all U.S. cities in the USCITYCOORDS table are selected that are south of Cairo, Egypt. The compound WHERE clause specifies the city of Cairo in the WORLDCITYCOORDS table and joins USCITYCOORDS and WORLDCITYCOORDS on their Latitude columns, using a less-than (`lt`) operator.

```
libname sql 'SAS-library';

proc sql;
   title 'US Cities South of Cairo, Egypt';
   select us.City, us.State, us.Latitude, world.city, world.latitude
      from sql.worldcitycoords world, sql.uscitycoords us
      where world.city = 'Cairo' and
            us.latitude lt world.latitude;
```

Output 3.6 *Using Comparison Operators to Join Tables*

US Cities South of Cairo, Egypt

City	State	Latitude	City	Latitude
Honolulu	HI	21	Cairo	30
Key West	FL	24	Cairo	30
Miami	FL	26	Cairo	30
San Antonio	TX	29	Cairo	30
Tampa	FL	28	Cairo	30

When you run this query, the following message is written to the SAS log:

Log 3.2 *Comparison Query Log Message*

```
NOTE: The execution of this query involves performing one or more Cartesian
      product joins that can not be optimized.
```

Recall that you see this message when you run a query that joins tables without specifying matching columns in a WHERE clause. PROC SQL also displays this message whenever tables are joined by using an inequality operator.

The Effects of Null Values on Joins

Most database products treat nulls as distinct entities and do not match them in joins. PROC SQL treats nulls as missing values and as matches for joins. Any null will match with any other null of the same type (character or numeric) in a join.

The following example joins Table One and Table Two on column B. There are null values in column B of both tables. Notice in the output that the null value in row c of Table One matches all the null values in Table Two. This is probably not the intended result for the join.

```
proc sql;
   title 'One and Two Joined';
   select one.a 'One', one.b, two.a 'Two', two.b
      from one, two
      where one.b=two.b;
```

Output 3.7 *Joining Tables That Contain Null Values*

Table One

a	b
a	1
b	2
c	.
d	4

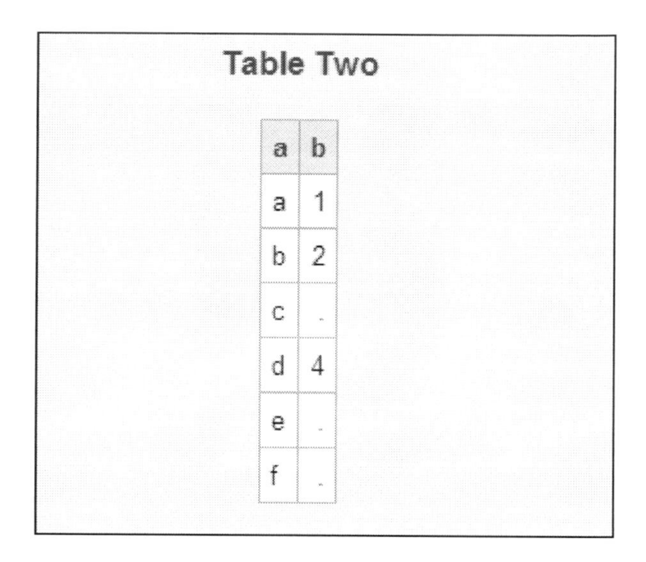

Table Two

a	b
a	1
b	2
c	.
d	4
e	.
f	.

One and Two Joined

One	b	Two	b
a	1	a	1
b	2	b	2
c	.	c	.
d	4	d	4
c	.	e	.
c	.	f	.

In order to specify only the nonmissing values for the join, use the IS NOT MISSING operator:

```
proc sql;
   select one.a 'One', one.b, two.a 'Two', two.b
      from one, two
      where one.b=two.b and
            one.b is not missing;
```

Output 3.8 *Results of Adding IS NOT MISSING to Joining Tables That Contain Null Values*

One and Two Joined

One	b	Two	b
a	1	a	1
b	2	b	2
d	4	d	4

Creating Multicolumn Joins

When a row is distinguished by a combination of values in more than one column, use all the necessary columns in the join. For example, a city name could exist in more than one country. To select the correct city, you must specify both the city and country columns in the joining query's WHERE clause.

This example displays the latitude and longitude of capital cities by joining the COUNTRIES table with the WORLDCITYCOORDS table. To minimize the number of rows in the example output, the first part of the WHERE expression selects capitals with names that begin with the letter `L` from the COUNTRIES table.

```
libname sql 'SAS library';

proc sql;
   title 'Coordinates of Capital Cities';
   select Capital format=$12., Name format=$12.,
          City format=$12., Country format=$12.,
          Latitude, Longitude
      from sql.countries, sql.worldcitycoords
      where Capital like 'L%' and
                Capital = City;
```

London occurs once as a capital city in the COUNTRIES table. However, in WORLDCITYCOORDS, London is found twice: as a city in England and again as a city in Canada. Specifying only **`Capital = City`** in the WHERE expression yields the following incorrect output:

Output 3.9 *Selecting Capital City Coordinates (incorrect output)*

Coordinates of Capital Cities

Capital	Name	City	Country	Latitude	Longitude
La Paz	Bolivia	La Paz	Bolivia	-16	-69
London	England	London	Canada	43	-81
Lima	Peru	Lima	Peru	-13	-77
Lisbon	Portugal	Lisbon	Portugal	39	-10
London	England	London	England	51	0

Notice in the output that the inner join incorrectly matches London, England, to both London, Canada, and London, England. By also joining the country name columns together (COUNTRIES.Name to WORLDCITYCOORDS.Country), the rows match correctly.

```
libname sql 'SAS-library';

proc sql;
   title 'Coordinates of Capital Cities';
   select Capital format=$12., Name format=$12.,
          City format=$12., Country format=$12.,
          latitude, longitude
      from sql.countries, sql.worldcitycoords
      where Capital like 'L%' and
            Capital = City and
            Name = Country;
```

Output 3.10 *Selecting Capital City Coordinates (correct output)*

Coordinates of Capital Cities

Capital	Name	City	Country	Latitude	Longitude
La Paz	Bolivia	La Paz	Bolivia	-16	-69
Lima	Peru	Lima	Peru	-13	-77
Lisbon	Portugal	Lisbon	Portugal	39	-10
London	England	London	England	51	0

Selecting Data from More than Two Tables

The data that you need could be located in more than two tables. For example, if you want to show the coordinates of the capitals of the states in the United States, then you need to join the UNITEDSTATES table, which contains the state capitals, with the USCITYCOORDS table, which contains the coordinates of cities in the United States. Because cities must be joined along with their states for an accurate join (similarly to the previous example), you must join the tables on both the city and state columns of the tables.

Joining the cities, by joining the UNITEDSTATES.Capital column to the USCITYCOORDS.City column, is straightforward. However, in the UNITEDSTATES table the Name column contains the full state name, while in USCITYCOORDS the states are specified by their postal code. It is therefore impossible to directly join the two tables on their state columns. To solve this problem, it is necessary to use the POSTALCODES table, which contains both the state names and their postal codes, as an intermediate table to make the correct relationship between UNITEDSTATES and USCITYCOORDS. The correct solution joins the UNITEDSTATES.Name column to the POSTALCODES.Name column (matching the full state names), and the POSTALCODES.Code column to the USCITYCOORDS.State column (matching the state postal codes).

```
libname sql 'SAS-library';

title 'Coordinates of State Capitals';
proc sql outobs=10;
```

```
select us.Capital format=$15., us.Name 'State' format=$15.,
       pc.Code, c.Latitude, c.Longitude
   from sql.unitedstates us, sql.postalcodes pc,
        sql.uscitycoords c
   where us.Capital = c.City and
         us.Name = pc.Name and
         pc.Code = c.State;
```

Output 3.11 *Selecting Data from More than Two Tables*

Coordinates of State Capitals

Capital	State	Code	Latitude	Longitude
Montgomery	Alabama	AL	32	-86
Juneau	Alaska	AK	58	-134
Phoenix	Arizona	AZ	33	-113
Little Rock	Arkansas	AR	35	-92
Sacramento	California	CA	38	-121
Denver	Colorado	CO	40	-105
Hartford	Connecticut	CT	42	-73
Dover	Delaware	DE	39	-76
Tallahassee	Florida	FL	31	-84
Atlanta	Georgia	GA	34	-84

Showing Relationships within a Single Table Using Self-Joins

When you need to show comparative relationships between values in a table, it is sometimes necessary to join columns within the same table. Joining a table to itself is called a self-join, or reflexive join. You can think of a self-join as PROC SQL making an internal copy of a table and joining the table to its copy.

For example, the following code uses a self-join to select cities that have average yearly high temperatures equal to the average yearly low temperatures of other cities.

```
libname sql 'SAS-library';

 proc sql;
    title "Cities' High Temps = Cities' Low Temps";
    select High.City format $12., High.Country format $12.,
           High.AvgHigh, ' | ',
           Low.City format $12., Low.Country format $12.,
           Low.AvgLow
       from sql.worldtemps High, sql.worldtemps Low
       where High.AvgHigh = Low.AvgLow and
             High.city ne Low.city and
             High.country ne Low.country;
```

Notice that the WORLDTEMPS table is assigned two aliases, **High** and **Low**. Conceptually, this makes a copy of the table so that a join can be made between the table and its copy. The WHERE clause selects those rows that have high temperature equal to low temperature.

The WHERE clause also prevents a city from being joined to itself (`City ne City` and `Country ne Country`), although, in this case, it is highly unlikely that the high temperature would be equal to the low temperature for the same city.

Output 3.12 *Joining a Table to Itself (Self-Join)*

Cities' High Temps = Cities' Low Temps

City	Country	AvgHigh		City	Country	AvgLow
Amsterdam	Netherlands	70	|	San Juan	Puerto Rico	70
Auckland	New Zealand	75	|	Lagos	Nigeria	75
Auckland	New Zealand	75	|	Manila	Philippines	75
Berlin	Germany	75	|	Lagos	Nigeria	75
Berlin	Germany	75	|	Manila	Philippines	75
Bogota	Colombia	69	|	Bangkok	Thailand	69
Cape Town	South Africa	70	|	San Juan	Puerto Rico	70
Copenhagen	Denmark	73	|	Singapore	Singapore	73
Dublin	Ireland	68	|	Bombay	India	68
Glasgow	Scotland	65	|	Nassau	Bahamas	65
London	England	73	|	Singapore	Singapore	73
Oslo	Norway	73	|	Singapore	Singapore	73
Reykjavik	Iceland	57	|	Caracas	Venezuela	57
Stockholm	Sweden	70	|	San Juan	Puerto Rico	70

Outer Joins

Overview of Outer Joins

Outer joins are inner joins that are augmented with rows from one table that do not match any row from the other table in the join. The resulting output includes rows that match and rows that do not match from the join's source tables. Nonmatching rows have null values in the columns from the unmatched table. Use the ON clause instead of the WHERE clause to specify the column or columns on which you are joining the tables. However, you can continue to use the WHERE clause to subset the query result.

Including Nonmatching Rows with the Left Outer Join

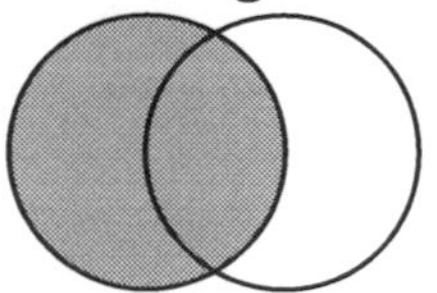

A left outer join lists matching rows and rows from the left-hand table (the first table listed in the FROM clause) that do not match any row in the right-hand table. A left join is specified with the keywords LEFT JOIN and ON.

For example, to list the coordinates of the capitals of international cities, join the COUNTRIES table, which contains capitals, with the WORLDCITYCOORDS table, which contains cities' coordinates, by using a left join. The left join lists all capitals, regardless of whether the cities exist in WORLDCITYCOORDS. Using an inner join would list only capital cities for which there is a matching city in WORLDCITYCOORDS.

```
libname sql 'SAS-library';

proc sql outobs=10;
   title 'Coordinates of Capital Cities';
   select Capital format=$20., Name 'Country' format=$20.,
          Latitude, Longitude
      from sql.countries a left join sql.worldcitycoords b
           on a.Capital = b.City and
              a.Name = b.Country
      order by Capital;
```

Output 3.13 *Left Join of COUNTRIES and WORLDCITYCOORDS*

Coordinates of Capital Cities

Capital	Country	Latitude	Longitude
	Channel Islands	.	.
Abu Dhabi	United Arab Emirates	.	.
Abuja	Nigeria	.	.
Accra	Ghana	5	0
Addis Ababa	Ethiopia	9	39
Algiers	Algeria	37	3
Almaty	Kazakhstan	.	.
Amman	Jordan	32	36
Amsterdam	Netherlands	52	5
Andorra la Vella	Andorra	.	.

Including Nonmatching Rows with the Right Outer Join

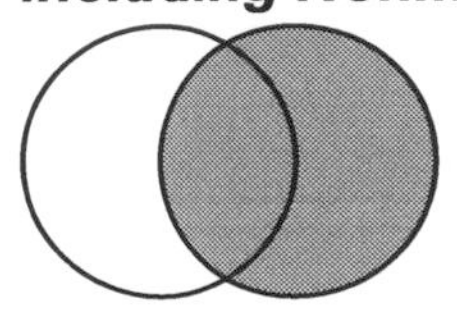

A right join, specified with the keywords RIGHT JOIN and ON, is the opposite of a left join: nonmatching rows from the right-hand table (the second table listed in the FROM clause) are included with all matching rows in the output. This example reverses the join of the last example; it uses a right join to select all the cities from the WORLDCITYCOORDS table and displays the population only if the city is the capital of a country (that is, if the city exists in the COUNTRIES table).

```
libname sql 'SAS-library';

proc sql outobs=10;
   title 'Populations of Capitals Only';
   select City format=$20., Country 'Country' format=$20.,
          Population
      from sql.countries right join sql.worldcitycoords
           on Capital = City and
              Name = Country
      order by City;
```

Output 3.14 *Right Join of COUNTRIES and WORLDCITYCOORDS*

Populations of Capitals Only

City	Country	Population
Abadan	Iran	.
Acapulco	Mexico	.
Accra	Ghana	17395511
Adana	Turkey	.
Addis Ababa	Ethiopia	59291170
Adelaide	Australia	.
Aden	Yemen	.
Ahmenabad	India	.
Algiers	Algeria	28171132
Alice Springs	Australia	.

Selecting All Rows with the Full Outer Join

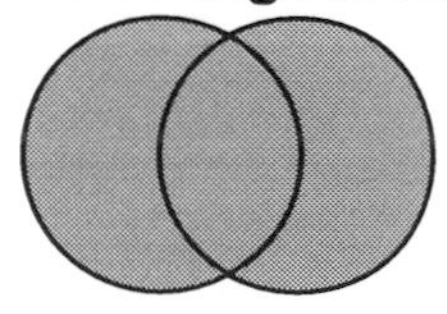

A full outer join, specified with the keywords FULL JOIN and ON, selects all matching and nonmatching rows. This example displays the first ten matching and nonmatching rows from the City and Capital columns of WORLDCITYCOORDS and COUNTRIES. Note that the pound sign (#) is used as a line split character in the labels.

```
libname sql 'SAS-library';

proc sql outobs=10;
   title 'Populations and/or Coordinates of World Cities';
   select City '#City#(WORLDCITYCOORDS)' format=$20.,
```

```
       Capital '#Capital#(COUNTRIES)' format=$20.,
       Population, Latitude, Longitude
   from sql.countries full join sql.worldcitycoords
        on Capital = City and
           Name = Country;
```

Output 3.15 *Full Outer Join of COUNTRIES and WORLDCITYCOORDS*

Populations and/or Coordinates of World Cities

City (WORLDCITYCOORDS)	Capital (COUNTRIES)	Population	Latitude	Longitude
		146436	.	.
Abadan		.	30	48
	Abu Dhabi	2818628	.	.
	Abuja	99062003	.	.
Acapulco		.	17	-100
Accra	Accra	17395511	5	0
Adana		.	37	35
Addis Ababa	Addis Ababa	59291170	9	39
Adelaide		.	-35	138
Aden		.	13	45

Specialty Joins

Overview of Specialty Joins

Three types of joins—cross joins, union joins, and natural joins—are special cases of the standard join types.

Including All Combinations of Rows with the Cross Join

A cross join is a Cartesian product; it returns the product of two tables. Like a Cartesian product, a cross join's output can be limited by a WHERE clause.

This example shows a cross join of the tables One and Two:

```
data one;
 input X Y $;
 datalines;
1 2
2 3
;


data two;
 input W Z $;
```

```
 datalines;
2 5
3 6
4 9
;
run;

proc sql;
  title 'Table One';
  select * from one;

  title 'Table Two';
  select * from two;

title;
quit;
```

Output 3.16 *Tables One and Two*

Table One

X	Y
1	2
2	3

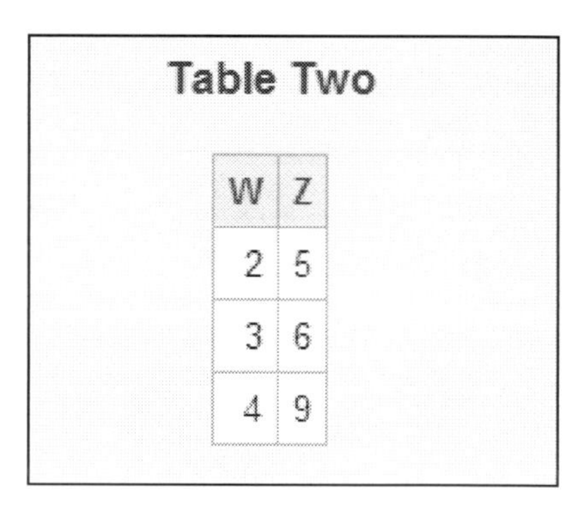

Table Two

W	Z
2	5
3	6
4	9

```
proc sql;
   title 'Table One and Table Two';
   select *
      from one cross join two;
```

Output 3.17 *Cross Join*

Table One and Table Two

X	Y	W	Z
1	2	2	5
1	2	3	6
1	2	4	9
2	3	2	5
2	3	3	6
2	3	4	9

Like a conventional Cartesian product, a cross join causes a note regarding Cartesian products in the SAS log.

Including All Rows with the Union Join

A union join combines two tables without attempting to match rows. All columns and rows from both tables are included. Combining tables with a union join is similar to combining them with the OUTER UNION set operator (see “Combining Queries with Set Operators” on page 102). A union join's output can be limited by a WHERE clause.

This example shows a union join of the same One and Two tables that were used earlier to demonstrate a cross join:

```
proc sql;
   select *
      from one union join two;
```

Output 3.18 *Union Join*

X	Y	W	Z
.		2	5
.		3	6
		4	9
1	2	.	
2	3		

Matching Rows with a Natural Join

A natural join automatically selects columns from each table to use in determining matching rows. With a natural join, PROC SQL identifies columns in each table that have the same name and type; rows in which the values of these columns are equal are returned as matching rows. The ON clause is implied.

This example produces the same results as the example in “Specifying the Order of Join Output” on page 77:

```
libname sql 'SAS-library';

proc sql outobs=6;
   title 'Oil Production/Reserves of Countries';
   select country, barrelsperday 'Production', barrels 'Reserve'
      from sql.oilprod natural join sql.oilrsrvs
      order by barrelsperday desc;
```

Output 3.19 *Natural Inner Join of OILPROD and OILRSRVS*

Oil Production/Reserves of Countries

Country	Production	Reserve
Saudi Arabia	9,000,000	260,000,000,000
United States of America	8,000,000	30,000,000,000
Iran	4,000,000	90,000,000,000
Norway	3,500,000	11,000,000,000
Mexico	3,400,000	50,000,000,000
China	3,000,000	25,000,000,000

The advantage of using a natural join is that the coding is streamlined. The ON clause is implied, and you do not need to use table aliases to qualify column names that are common to both tables. These two queries return the same results:

```
proc sql;
   select a.W, a.X, Y, Z
   from table1 a left join table2 b
   on a.W=b.W and a.X=b.X
   order by a.W;
```

```
proc sql;
   select W, X, Y, Z
   from table1 natural left join table2
   order by W;
```

If you specify a natural join on tables that do not have at least one column with a common name and type, then the result is a Cartesian product. You can use a WHERE clause to limit the output.

Because the natural join makes certain assumptions about what you want to accomplish, you should know your data thoroughly before using it. You could get unexpected or incorrect results. For example, if you are expecting two tables to have only one column in common when they actually have two. You can use the FEEDBACK option to see exactly how PROC SQL is implementing your query. See “Using PROC SQL Options to Create and Debug Queries” on page 136 for more information about the FEEDBACK option.

A natural join assumes that you want to base the join on equal values of all pairs of common columns. To base the join on inequalities or other comparison operators, use standard inner or outer join syntax.

Using the Coalesce Function in Joins

As you can see from the previous examples, the nonmatching rows in outer joins contain missing values. By using the COALESCE function, you can overlay columns so that only the row from the table that contains data is listed. Recall that COALESCE takes a list of columns as its arguments and returns the first nonmissing value that it encounters.

This example adds the COALESCE function to the previous example to overlay the COUNTRIES.Capital, WORLDCITYCOORDS.City, and COUNTRIES.Name columns.

COUNTRIES.Name is supplied as an argument to COALESCE because some islands do not have capitals.

```
libname sql 'SAS-library';

proc sql outobs=10;
   title 'Populations and/or Coordinates of World Cities';
   select coalesce(Capital, City,Name)format=$20. 'City',
          coalesce(Name, Country) format=$20. 'Country',
          Population, Latitude, Longitude
      from sql.countries full join sql.worldcitycoords
           on Capital = City and
           Name = Country;
```

Output 3.20 *Using COALESCE in Full Outer Join of COUNTRIES and WORLDCITYCOORDS*

Populations and/or Coordinates of World Cities

City	Country	Population	Latitude	Longitude
Channel Islands	Channel Islands	146436	.	.
Abadan	Iran	.	30	48
Abu Dhabi	United Arab Emirates	2818628	.	.
Abuja	Nigeria	99062003	.	.
Acapulco	Mexico	.	17	-100
Accra	Ghana	17395511	5	0
Adana	Turkey	.	37	35
Addis Ababa	Ethiopia	59291170	9	39
Adelaide	Australia	.	-35	138
Aden	Yemen	.	13	45

COALESCE can be used in both inner and outer joins. For more information about COALESCE, see "Replacing Missing Values" on page 35.

Comparing DATA Step Match-Merges with PROC SQL Joins

Overview of Comparing DATA Step Match-Merges with PROC SQL Joins

Many SAS users are familiar with using a DATA step to merge data sets. This section compares merges to joins. DATA step match-merges and PROC SQL joins can produce the same results. However, a significant difference between a match-merge and a join is that you do not have to sort the tables before you join them.

When All of the Values Match

When all of the values match in the BY variable and there are no duplicate BY variables, you can use an inner join to produce the same result as a match-merge. To demonstrate this result, here are two tables that have the column Flight in common. The values of Flight are the same in both tables:

```
FLTSUPER                      FLTDEST

Flight  Supervisor            Flight  Destination

  145   Kang                    145   Brussels
  150   Miller                  150   Paris
  155   Evanko                  155   Honolulu
```

FLTSUPER and FLTDEST are already sorted by the matching column Flight. A DATA step merge produces Output 3.21 on page 92.

```
data fltsuper;
input Flight Supervisor $;
datalines;
145   Kang
150   Miller
155   Evanko
;
data fltdest;
input Flight Destination $;
datalines;
145   Brussels
150   Paris
155   Honolulu
;
run;

data merged;
   merge FltSuper FltDest;
   by Flight;
run;

proc print data=merged noobs;
   title 'Table MERGED';
run;
```

Output 3.21 *Merged Tables When All the Values Match*

Table MERGED

Flight	Supervisor	Destination
145	Kang	Brussels
150	Miller	Paris
155	Evanko	Honolulu

With PROC SQL, presorting the data is not necessary. The following PROC SQL join gives the same result as that shown in Output 3.21 on page 92.

```
proc sql;
   title 'Table MERGED';
   select s.flight, Supervisor, Destination
```

```
      from fltsuper s, fltdest d
      where s.Flight=d.Flight;
```

When Only Some of the Values Match

When only some of the values match in the BY variable, you can use an outer join to produce the same result as a match-merge. To demonstrate this result, here are two tables that have the column Flight in common. The values of Flight are not the same in both tables:

```
FLTSUPER                      FLTDEST

Flight  Supervisor            Flight  Destination

  145   Kang                    145   Brussels
  150   Miller                  150   Paris
  155   Evanko                  165   Seattle
  157   Lei
```

A DATA step merge produces Output 3.22 on page 93:

```
data merged;
   merge fltsuper fltdest;
   by flight;
run;
proc print data=merged noobs;
   title 'Table MERGED';
run;
```

Output 3.22 *Merged Tables When Some of the Values Match*

Table MERGED

Flight	Supervisor	Destination
145	Kang	Brussels
150	Miller	Paris
155	Evanko	
157	Lei	
165		Seattle

To get the same result with PROC SQL, use an outer join so that the query result will contain the nonmatching rows from the two tables. In addition, use the COALESCE function to overlay the Flight columns from both tables. The following PROC SQL join gives the same result as that shown in Output 3.22 on page 93:

```
proc sql;
   select coalesce(s.Flight,d.Flight) as Flight, Supervisor, Destination
      from fltsuper s full join fltdest d
         on s.Flight=d.Flight;
```

When the Position of the Values Is Important

When you want to merge two tables and the position of the values is important, you might need to use a DATA step merge. To demonstrate this idea, here are two tables to consider:

```
FLTSUPER                          FLTDEST

Flight  Supervisor              Flight  Destination

  145   Kang                     145    Brussels
  145   Ramirez                  145    Edmonton
  150   Miller                   150    Paris
  150   Picard                   150    Madrid
  155   Evanko                   165    Seattle
  157   Lei
```

For Flight 145, **Kang** matches with **Brussels** and **Ramirez** matches with **Edmonton**. Because the DATA step merges data based on the position of values in BY groups, the values of Supervisor and Destination match appropriately. A DATA step merge produces Output 3.23 on page 94:

```
data merged;
   merge fltsuper fltdest;
   by flight;
run;
proc print data=merged noobs;
   title 'Table MERGED';
run;
```

Output 3.23 *Match-Merge of the FLTSUPER and FLTDEST Tables*

Table MERGED

Flight	Supervisor	Destination
145	Kang	Brussels
145	Ramirez	Edmonton
150	Miller	Paris
150	Picard	Madrid
155	Evanko	
157	Lei	
165		Seattle

PROC SQL does not process joins according to the position of values in BY groups. Instead, PROC SQL processes data only according to the data values. Here is the result of an inner join for FLTSUPER and FLTDEST:

```
proc sql;
   title 'Table JOINED';
   select *
```

```
from fltsuper s, fltdest d
where s.Flight=d.Flight;
```

Output 3.24 *PROC SQL Join of the FLTSUPER and FLTDEST Tables*

Table JOINED

Flight	Supervisor	Flight	Destination
145	Kang	145	Brussels
145	Kang	145	Edmonton
145	Ramirez	145	Brussels
145	Ramirez	145	Edmonton
150	Miller	150	Paris
150	Miller	150	Madrid
150	Picard	150	Paris
150	Picard	150	Madrid

PROC SQL builds the Cartesian product and then lists the rows that meet the WHERE clause condition. The WHERE clause returns two rows for each supervisor, one row for each destination. Because Flight has duplicate values and there is no other matching column, there is no way to associate **`Kang`** only with **`Brussels`**, **`Ramirez`** only with **`Edmonton`**, and so on.

For more information about DATA step match-merges, see *SAS Statements: Reference*.

Using Subqueries to Select Data

While a table join combines multiple tables into a new table, a subquery (enclosed in parentheses) selects rows from one table based on values in another table. A subquery, or inner query, is a query expression that is nested as part of another query expression. Depending on the clause that contains it, a subquery can return a single value or multiple values. Subqueries are most often used in the WHERE and the HAVING expressions.

Single-Value Subqueries

A single-value subquery returns a single row and column. It can be used in a WHERE or HAVING clause with a comparison operator. The subquery must return only one value, or else the query fails and an error message is printed to the log.

This query uses a subquery in its WHERE clause to select U.S. states that have a population greater than Belgium. The subquery is evaluated first, and then it returns the population of Belgium to the outer query.

```
libname sql 'SAS-library';

proc sql;
   title 'U.S. States with Population Greater than Belgium';
```

```
select Name 'State' , population format=comma10.
   from sql.unitedstates
   where population gt
            (select population from sql.countries
               where name = "Belgium");
```

Internally, this is what the query looks like after the subquery has executed:

```
proc sql;
   title 'U.S. States with Population Greater than Belgium';
   select Name 'State', population format=comma10.
      from sql.unitedstates
      where population gt 10162614;
```

The outer query lists the states whose populations are greater than the population of Belgium.

Output 3.25 *Single-Value Subquery*

U.S. States with Population Greater than Belgium

State	Population
California	31,518,948
Florida	13,814,408
Illinois	11,813,091
New York	18,377,334
Ohio	11,200,790
Pennsylvania	12,167,566
Texas	18,209,994

Multiple-Value Subqueries

A multiple-value subquery can return more than one value from one column. It is used in a WHERE or HAVING expression that contains IN or a comparison operator that is modified by ANY or ALL. This example displays the populations of oil-producing countries. The subquery first returns all countries that are found in the OILPROD table. The outer query then matches countries in the COUNTRIES table to the results of the subquery.

```
libname sql 'SAS-library';

proc sql outobs=5;
   title 'Populations of Major Oil Producing Countries';
   select name 'Country', Population format=comma15.
      from sql.countries
      where Name in
            (select Country from sql.oilprod);
```

Output 3.26 *Multiple-Value Subquery Using IN*

Populations of Major Oil Producing Countries

Country	Population
Algeria	28,171,132
Canada	28,392,302
China	1,202,215,077
Egypt	59,912,259
Indonesia	202,393,859

If you use the NOT IN operator in this query, then the query result will contain all the countries that are not contained in the OILPROD table.

```
libname sql 'SAS-library';

proc sql outobs=5;
   title 'Populations of NonMajor Oil Producing Countries';
   select name 'Country', Population format=comma15.
      from sql.countries
      where Name not in
            (select Country from sql.oilprod);
```

Output 3.27 *Multiple-Value Subquery Using NOT IN*

Populations of NonMajor Oil Producing Countries

Country	Population
Afghanistan	17,070,323
Albania	3,407,400
Andorra	64,634
Angola	9,901,050
Antigua and Barbuda	65,644

Correlated Subqueries

The previous subqueries have been simple subqueries that are self-contained and that execute independently of the outer query. A correlated subquery requires a value or values to be passed to it by the outer query. After the subquery runs, it passes the results back to the outer query. Correlated subqueries can return single or multiple values.

This example selects all major oil reserves of countries on the continent of Africa.

```
libname sql 'SAS-library';

proc sql;
```

```
title 'Oil Reserves of Countries in Africa';
select * from sql.oilrsrvs o
   where 'Africa' =
          (select Continent from sql.countries c
            where c.Name = o.Country);
```

The outer query selects the first row from the OILRSRVS table and then passes the value of the Country column, **Algeria**, to the subquery. At this point, the subquery internally looks like this:

```
(select Continent from sql.countries c
        where c.Name = 'Algeria');
```

The subquery selects that country from the COUNTRIES table. The subquery then passes the country's continent back to the WHERE clause in the outer query. If the continent is Africa, then the country is selected and displayed. The outer query then selects each subsequent row from the OILRSRVS table and passes the individual values of Country to the subquery. The subquery returns the appropriate values of Continent to the outer query for comparison in its WHERE clause.

Note that the WHERE clause uses an = (equal) operator. You can use an = (equal) operator if the subquery returns only a single value. However, if the subquery returns multiple values, then you must use IN or a comparison operator with ANY or ALL. For detailed information about the operators that are available for use with subqueries, see Chapter 7, “SQL Procedure,” on page 209.

Output 3.28 *Correlated Subquery*

Oil Reserves of Countries in Africa

Country	Barrels
Algeria	9,200,000,000
Egypt	4,000,000,000
Gabon	1,000,000,000
Libya	30,000,000,000
Nigeria	16,000,000,000

Testing for the Existence of a Group of Values

The EXISTS condition tests for the existence of a set of values. An EXISTS condition is true if any rows are produced by the subquery, and it is false if no rows are produced. Conversely, the NOT EXISTS condition is true when a subquery produces an empty table.

This example produces the same result as Output 3.28 on page 98. EXISTS checks for the existence of countries that have oil reserves on the continent of Africa. Note that the WHERE clause in the subquery now contains the condition **Continent = 'Africa'** that was in the outer query in the previous example.

```
libname sql 'SAS-library';

proc sql;
```

```
title 'Oil Reserves of Countries in Africa';
select * from sql.oilrsrvs o
   where exists
          (select Continent from sql.countries c
             where o.Country = c.Name and
                   Continent = 'Africa');
```

Output 3.29 *Testing for the Existence of a Group of Values*

Oil Reserves of Countries in Africa

Country	Barrels
Algeria	9,200,000,000
Egypt	4,000,000,000
Gabon	1,000,000,000
Libya	30,000,000,000
Nigeria	16,000,000,000

Multiple Levels of Subquery Nesting

Subqueries can be nested so that the innermost subquery returns a value or values to be used by the next outer query. Then, that subquery's value or values are used by the next outer query, and so on. Evaluation always begins with the innermost subquery and works outward.

This example lists cities in Africa that are in countries with major oil reserves.

1. The innermost query is evaluated first. It returns countries that are located on the continent of Africa.
2. The outer subquery is evaluated. It returns a subset of African countries that have major oil reserves by comparing the list of countries that was returned by the inner subquery against the countries in OILRSRVS.
3. Finally, the WHERE clause in the outer query lists the coordinates of the cities that exist in the WORLDCITYCOORDS table whose countries match the results of the outer subquery.

```
libname sql 'SAS-library';

proc sql;
   title 'Coordinates of African Cities with Major Oil Reserves';
   select * from sql.worldcitycoords
3    where country in
 2      (select Country from sql.oilrsrvs o
           where o.Country in
        1 (select Name from sql.countries c
             where c.Continent='Africa'));
```

Output 3.30 *Multiple Levels of Subquery Nesting*

Coordinates of African Cities with Major Oil Reserves

City	Country	Latitude	Longitude
Algiers	Algeria	37	3
Cairo	Egypt	30	31
Benghazi	Libya	33	21
Lagos	Nigeria	6	3

Combining a Join with a Subquery

You can combine joins and subqueries in a single query. Suppose that you want to find the city nearest to each city in the USCITYCOORDS table. The query must first select a city **A**, compute the distance from a city **A** to every other city, and finally select the city with the minimum distance from city A. This can be done by joining the USCITYCOORDS table to itself (self-join) and then determining the closest distance between cities by using another self-join in a subquery.

This is the formula to determine the distance between coordinates:

```
SQRT(((Latitude2-Latitude1)**2) + ((Longitude2-Longitude1)**2))
```

Although the results of this formula are not exactly accurate because of the distortions caused by the curvature of the earth, they are accurate enough for this example to determine whether one city is closer than another.

```
libname sql 'SAS-library';

proc sql outobs=10;
   title 'Neighboring Cities';
   select a.City format=$10., a.State,
          a.Latitude 'Lat', a.Longitude 'Long',
          b.City format=$10., b.State,
          b.Latitude 'Lat', b.Longitude 'Long',
          sqrt(((b.latitude-a.latitude)**2) +
               ((b.longitude-a.longitude)**2)) as dist format=6.1
      from sql.uscitycoords a, sql.uscitycoords b
      where a.city ne b.city and
            calculated dist =
            (select min(sqrt(((d.latitude-c.latitude)**2) +
                             ((d.longitude-c.longitude)**2)))
                from sql.uscitycoords c, sql.uscitycoords d
                where c.city = a.city and
                      c.state = a.state and
                      d.city ne c.city)
            order by a.city;
```

Output 3.31 *Combining a Join with a Subquery*

Neighboring Cities

City	State	Lat	Long	City	State	Lat	Long	dist
Albany	NY	43	-74	Hartford	CT	42	-73	1.4
Albuquerqu	NM	36	-106	Santa Fe	NM	36	-106	0.0
Amarillo	TX	35	-102	Carlsbad	NM	32	-104	3.6
Anchorage	AK	61	-150	Nome	AK	64	-165	15.3
Annapolis	MD	39	-77	Washington	DC	39	-77	0.0
Atlanta	GA	34	-84	Knoxville	TN	36	-84	2.0
Augusta	ME	44	-70	Portland	ME	44	-70	0.0
Austin	TX	30	-98	San Antoni	TX	29	-98	1.0
Baker	OR	45	-118	Lewiston	ID	46	-117	1.4
Baltimore	MD	39	-76	Dover	DE	39	-76	0.0

The outer query joins the table to itself and determines the distance between the first city A1 in table A and city B2 (the first city that is not equal to city A1) in Table B. PROC SQL then runs the subquery. The subquery does another self-join and calculates the minimum distance between city A1 and all other cities in the table other than city A1. The outer query tests to see whether the distance between cities A1 and B2 is equal to the minimum distance that was calculated by the subquery. If they are equal, then a row that contains cities A1 and B2 with their coordinates and distance is written.

When to Use Joins and Subqueries

Use a join or a subquery any time that you reference information from multiple tables. Joins and subqueries are often used together in the same query. In many cases, you can solve a data retrieval problem by using a join, a subquery, or both. Here are some guidelines for using joins and queries.

- If your report needs data that is from more than one table, then you must perform a join. Whenever multiple tables (or views) are listed in the FROM clause, those tables become joined.
- If you need to combine related information from different rows within a table, then you can join the table with itself.
- Use subqueries when the result that you want requires more than one query and each subquery provides a subset of the table involved in the query.
- If a membership question is asked, then a subquery is usually used. If the query requires a NOT EXISTS condition, then you must use a subquery because NOT EXISTS operates only in a subquery; the same principle holds true for the EXISTS condition.

- Many queries can be formulated as joins or subqueries. Although the PROC SQL query optimizer changes some subqueries to joins, a join is generally more efficient to process.

Combining Queries with Set Operators

Working with Two or More Query Results

PROC SQL can combine the results of two or more queries in various ways by using the following set operators:

UNION
: produces all unique rows from both queries.

EXCEPT
: produces rows that are part of the first query only.

INTERSECT
: produces rows that are common to both query results.

OUTER UNION
: concatenates the query results.

The operator is used between the two queries, for example:

```
select columns from table
set-operator
select columns from table;
```

Place a semicolon after the last SELECT statement only. Set operators combine columns from two queries based on their position in the referenced tables without regard to the individual column names. Columns in the same relative position in the two queries must have the same data types. The column names of the tables in the first query become the column names of the output table. For information about using set operators with more than two query results, see the Chapter 7, "SQL Procedure," on page 209. The following optional keywords give you more control over set operations:

ALL
: does not suppress duplicate rows. When the keyword ALL is specified, PROC SQL does not make a second pass through the data to eliminate duplicate rows. Thus, using ALL is more efficient than not using it. ALL is not allowed with the OUTER UNION operator.

CORRESPONDING (CORR)
: overlays columns that have the same name in both tables. When used with EXCEPT, INTERSECT, and UNION, CORR suppresses columns that are not in both tables.

Each set operator is described and used in an example based on the following two tables.

Output 3.32 *Tables Used in Set Operation Examples*

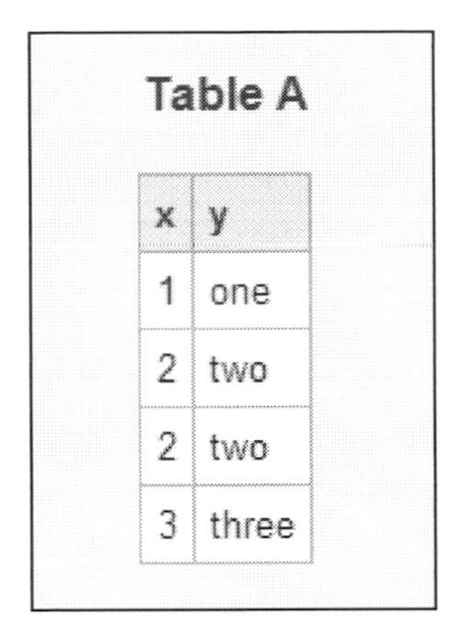

Table A

x	y
1	one
2	two
2	two
3	three

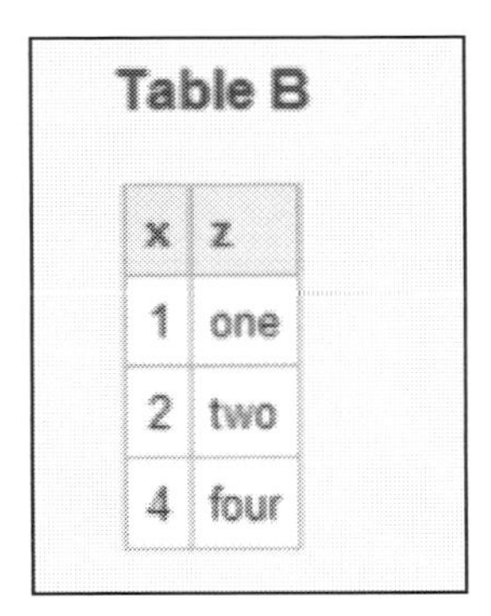

Table B

x	z
1	one
2	two
4	four

Whereas join operations combine tables horizontally, set operations combine tables vertically. Therefore, the set diagrams that are included in each section are displayed vertically.

Producing Unique Rows from Both Queries (UNION)

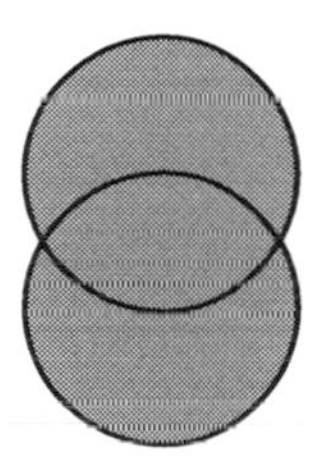

The UNION operator combines two query results. It produces all the unique rows that result from both queries. That is, it returns a row if it occurs in the first table, the second, or both. UNION does not return duplicate rows. If a row occurs more than once, then only one occurrence is returned.

```
proc sql;
   title 'A UNION B';
   select * from sql.a
   union
   select * from sql.b;
```

Output 3.33 *Producing Unique Rows from Both Queries (UNION)*

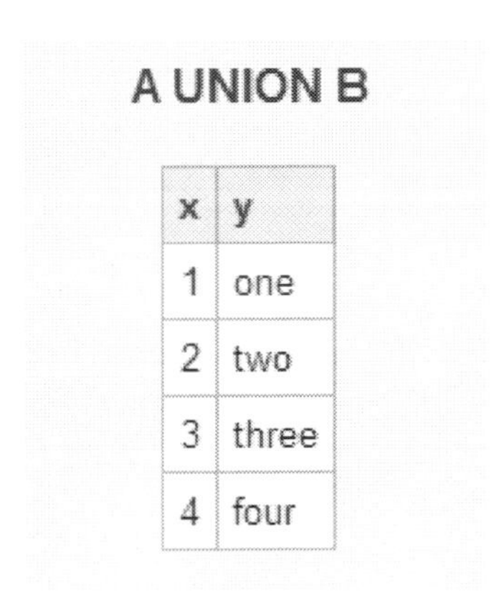

A UNION B

x	y
1	one
2	two
3	three
4	four

You can use the ALL keyword to request that duplicate rows remain in the output.

```
proc sql;
   title 'A UNION ALL B';
   select * from sql.a
   union all
   select * from sql.b;
```

Output 3.34 *Producing Rows from Both Queries (UNION ALL)*

A UNION ALL B

x	y
1	one
2	two
2	two
3	three
1	one
2	two
4	four

Producing Rows That Are in Only the First Query Result (EXCEPT)

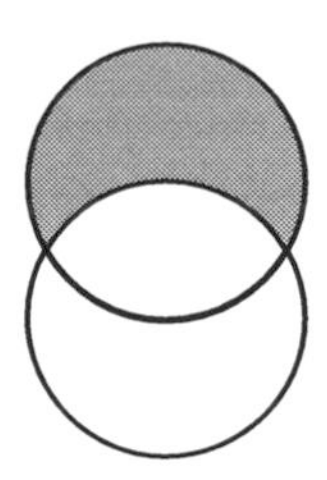

The EXCEPT operator returns rows that result from the first query but not from the second query. In this example, the row that contains the values `3` and **`three`** exists in the first query (table A) only and is returned by EXCEPT.

```
proc sql;
   title 'A EXCEPT B';
   select * from sql.a
   except
   select * from sql.b;
```

Output 3.35 *Producing Rows That Are in Only the First Query Result (EXCEPT)*

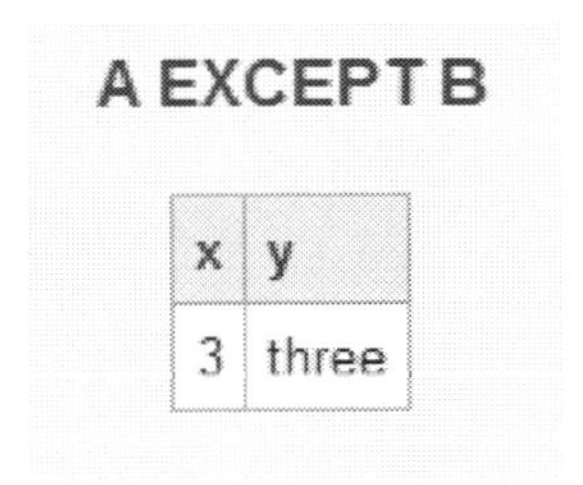

A EXCEPT B

x	y
3	three

Note that the duplicated row in Table A containing the values **2** and **two** does not appear in the output. EXCEPT does not return duplicate rows that are unmatched by rows in the second query. Adding ALL keeps any duplicate rows that do not occur in the second query.

```
proc sql;
   title 'A EXCEPT ALL B';
   select * from sql.a
   except all
   select * from sql.b;
```

Output 3.36 *Producing Rows That Are in Only the First Query Result (EXCEPT ALL)*

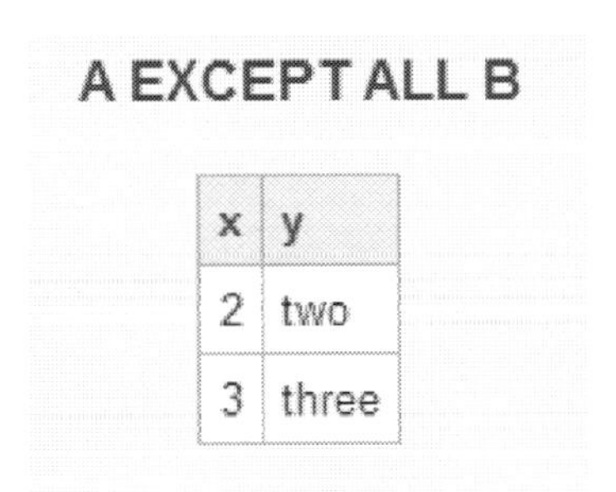

A EXCEPT ALL B

x	y
2	two
3	three

Producing Rows That Belong to Both Query Results (INTERSECT)

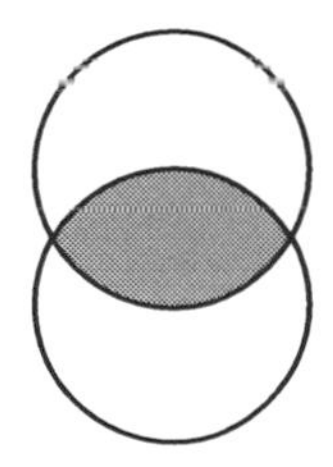

The INTERSECT operator returns rows from the first query that also occur in the second.

```
proc sql;
   title 'A INTERSECT B';
   select * from sql.a
   intersect
   select * from sql.b;
```

Output 3.37 *Producing Rows That Belong to Both Query Results (INTERSECT)*

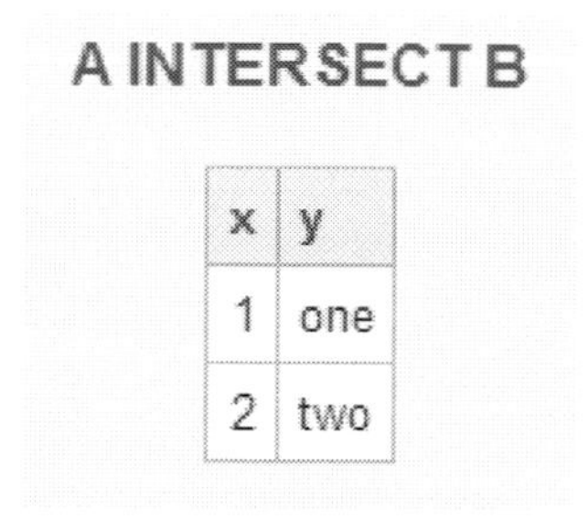
A INTERSECT B

x	y
1	one
2	two

The output of an INTERSECT ALL operation contains the rows produced by the first query that are matched one-to-one with a row produced by the second query. In this example, the output of INTERSECT ALL is the same as INTERSECT.

Concatenating Query Results (OUTER UNION)

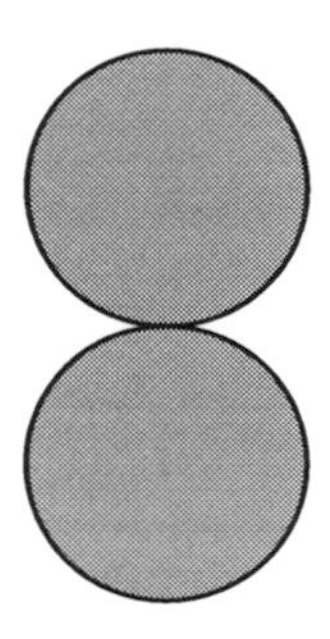

The OUTER UNION operator concatenates the results of the queries. This example concatenates tables A and B.

```
proc sql;
   title 'A OUTER UNION B';
   select * from sql.a
   outer union
   select * from sql.b;
```

Output 3.38 *Concatenating the Query Results (OUTER UNION)*

A OUTER UNION B

x	y	x	z
1	one	.	
2	two	.	
2	two	.	
3	three	.	
.		1	one
.		2	two
.		4	four

Notice that OUTER UNION does not overlay columns from the two tables. To overlay columns in the same position, use the CORRESPONDING keyword.

```
proc sql;
   title 'A OUTER UNION CORR B';
   select * from sql.a
   outer union corr
   select * from sql.b;
```

Output 3.39 *Concatenating the Query Results (OUTER UNION CORR)*

A OUTER UNION CORR B

x	y	z
1	one	
2	two	
2	two	
3	three	
1		one
2		two
4		four

Producing Rows from the First Query or the Second Query

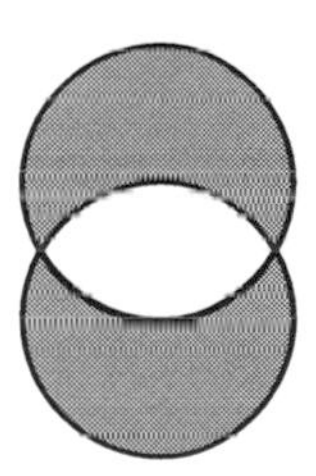

There is no keyword in PROC SQL that returns unique rows from the first and second table, but not rows that occur in both. Here is one way that you can simulate this operation:

```
(query1 except query2)
union
(query2 except query1)
```

This example shows how to use this operation.

```
proc sql;
   title 'A EXCLUSIVE UNION B';
   (select * from sql.a
         except
         select * from sql.b)
   union
   (select * from sql.b
```

```
    except
    select * from sql.a);
```

Output 3.40 *Producing Rows from the First Query or the Second Query*

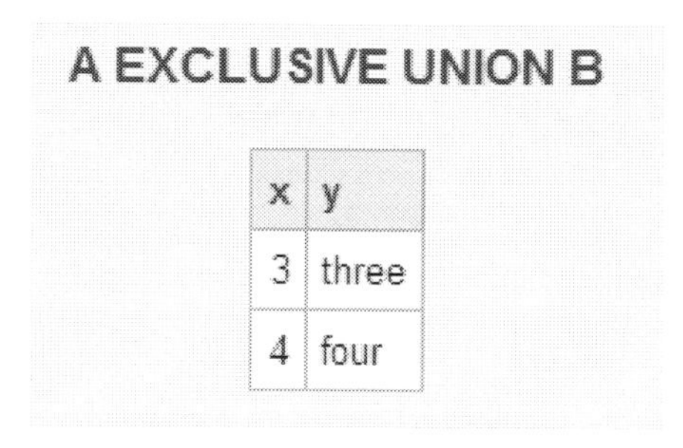
A EXCLUSIVE UNION B

x	y
3	three
4	four

The first EXCEPT returns one unique row from the first table (table A) only. The second EXCEPT returns one unique row from the second table (table B) only. The middle UNION combines the two results. Thus, this query returns the row from the first table that is not in the second table, as well as the row from the second table that is not in the first table.

Chapter 4

Creating and Updating Tables and Views

Introduction

This chapter shows you how to perform the following tasks:

- create a table
- update tables
- alter existing tables
- delete a table
- create indexes
- use integrity constraints in table creation
- create views

Creating Tables

The CREATE TABLE statement enables you to create tables without rows from column definitions or to create tables from a query result. You can also use CREATE TABLE to copy an existing table.

Creating Tables from Column Definitions

You can create a new table without rows by using the CREATE TABLE statement to define the columns and their attributes. You can specify a column's name, type, length, informat, format, and label.

The following CREATE TABLE statement creates the NEWSTATES table:

```
proc sql;
   create table sql.newstates
          (state char(2),            /* 2-character column for          */
                                     /* state abbreviation              */

           date num                  /* column for date of entry into the US */
                informat=date9.      /* with an informat                */
                format=date9.,       /* and format of DATE9.            */

           population num);          /* column for population           */
```

The table NEWSTATES has three columns and 0 rows. The char(2) modifier is used to change the length for State.

Use the DESCRIBE TABLE statement to verify that the table exists and to see the column attributes. The following DESCRIBE TABLE statement writes a CREATE TABLE statement to the SAS log:

```
proc sql;
   describe table sql.newstates;
```

Log 4.1 *Table Created from Column Definitions*

```
1  proc sql;
2     describe table sql.newstates;
NOTE: SQL table SQL.NEWSTATES was created like:

create table SQL.NEWSTATES( bufsize=8192 )
   (
    state char(2),
    date num format=DATE9. informat=DATE9.,
    population num
   );
```

DESCRIBE TABLE writes a CREATE TABLE statement to the SAS log even if you did not create the table with the CREATE TABLE statement. You can also use the CONTENTS statement in the DATASETS procedure to get a description of NEWSTATES.

Creating Tables from a Query Result

To create a PROC SQL table from a query result, use a CREATE TABLE statement, and place it before the SELECT statement. When a table is created this way, its data is derived from the table or view that is referenced in the query's FROM clause. The new table's column names are as specified in the query's SELECT clause list. The column attributes (the type, length, informat, and format) are the same as those of the selected source columns.

The following CREATE TABLE statement creates the DENSITIES table from the COUNTRIES table. The newly created table is not displayed in SAS output unless you query the table. Note the use of the OUTOBS option, which limits the size of the DENSITIES table to 10 rows.

```
libname sql 'SAS-library';

proc sql outobs=10;
   title 'Densities of Countries';
   create table sql.densities as
      select Name 'Country' format $15.,
             Population format=comma10.0,
             Area as SquareMiles,
             Population/Area format=6.2 as Density
         from sql.countries;

   select * from sql.densities;
```

Output 4.1 *Table Created from a Query Result*

Densities of Countries

Country	Population	SquareMiles	Density
Afghanistan	17,070,323	251825	67.79
Albania	3,407,400	11100	306.97
Algeria	28,171,132	919595	30.63
Andorra	64,634	200	323.17
Angola	9,901,050	481300	20.57
Antigua and Bar	65,644	171	383.88
Argentina	34,248,705	1073518	31.90
Armenia	3,556,864	11500	309.29
Australia	18,255,944	2966200	6.15
Austria	8,033,746	32400	247.96

The following DESCRIBE TABLE statement writes a CREATE TABLE statement to the SAS log:

```
proc sql;
   describe table sql.densities;
```

Log 4.2 *SAS Log for DESCRIBE TABLE Statement for DENSITIES*

```
NOTE: SQL table SQL.DENSITIES was created like:

create table SQL.DENSITIES( bufsize=8192 )
  (
   Name char(35) format=$15. informat=$35. label='Country',
   Population num format=COMMA10. informat=BEST8. label='Population',
   SquareMiles num format=BEST8. informat=BEST8. label='SquareMiles',
   Density num format=6.2
  );
```

In this form of the CREATE TABLE statement, assigning an alias to a column renames the column, while assigning a label does not. In this example, the Area column has been renamed to SquareMiles, and the calculated column has been named Densities. However, the Name column retains its name, and its display label is `Country`.

Creating Tables like an Existing Table

To create an empty table that has the same columns and attributes as an existing table or view, use the LIKE clause in the CREATE TABLE statement. In the following example, the CREATE TABLE statement creates the NEWCOUNTRIES table with six columns and 0 rows and with the same column attributes as those in COUNTRIES. The DESCRIBE TABLE statement writes a CREATE TABLE statement to the SAS log:

```
proc sql;
   create table sql.newcountries
      like sql.countries;

   describe table sql.newcountries;
```

Log 4.3 *SAS Log for DESCRIBE TABLE Statement for NEWCOUNTRIES*

```
NOTE: SQL table SQL.NEWCOUNTRIES was created like:

create table SQL.NEWCOUNTRIES( bufsize=16384 )
  (
   Name char(35) format=$35. informat=$35.,
   Capital char(35) format=$35. informat=$35. label='Capital',
   Population num format=BEST8. informat=BEST8. label='Population',
   Area num format=BEST8. informat=BEST8.,
   Continent char(35) format=$35. informat=$35. label='Continent',
   UNDate num format=YEAR4.
  );
```

Copying an Existing Table

A quick way to copy a table using PROC SQL is to use the CREATE TABLE statement with a query that returns an entire table. This example creates COUNTRIES1, which contains a copy of all the columns and rows that are in COUNTRIES:

```
create table countries1 as
   select * from sql.countries;
```

Using Data Set Options

You can use SAS data set options in the CREATE TABLE statement. The following CREATE TABLE statement creates COUNTRIES2 from COUNTRIES. The DROP= option deletes the UNDate column, and UNDate does not become part of COUNTRIES2:

```
create table countries2 as
   select * from sql.countries(drop=UNDate);
```

Inserting Rows into Tables

Use the INSERT statement to insert data values into tables. The INSERT statement first adds a new row to an existing table, and then inserts the values that you specify into the row. You specify values by using a SET clause or VALUES clause. You can also insert the rows resulting from a query. Under most conditions, you can insert data into tables through PROC SQL and SAS/ACCESS views. For more information, see "Updating a View" on page 130.

Inserting Rows with the SET Clause

With the SET clause, you assign values to columns by name. The columns can appear in any order in the SET clause. The following INSERT statement uses multiple SET clauses to add two rows to NEWCOUNTRIES:

```
libname sql 'SAS-library';

proc sql;
create table sql.newcountries
   like sql.countries;

proc sql;
title "World's Largest Countries";
   insert into sql.newcountries
   select * from sql.countries
      where population ge 130000000;

proc sql;
   insert into sql.newcountries
      set name='Bangladesh',
          capital='Dhaka',
          population=126391060
      set name='Japan',
          capital='Tokyo',
          population=126352003;

   title "World's Largest Countries";
   select name format=$20.,
          capital format=$15.,
          population format=comma15.0
      from sql.newcountries;
```

Output 4.2 *Rows Inserted with the SET Clause*

World's Largest Countries

Name	Capital	Population
Brazil	Brasilia	160,310,357
China	Beijing	1,202,215,077
India	New Delhi	929,009,120
Indonesia	Jakarta	202,393,859
Russia	Moscow	151,089,979
United States	Washington	263,294,808
Bangladesh	Dhaka	126,391,060
Japan	Tokyo	126,352,003

Note the following features of SET clauses:

- As with other SQL clauses, use commas to separate columns. In addition, you must use a semicolon after the last SET clause only.
- If you omit data for a column, then the value in that column is a missing value.
- To specify that a value is missing, use a blank in single quotation marks for character values and a period for numeric values.

Inserting Rows with the VALUES Clause

With the VALUES clause, you assign values to a column by position. The following INSERT statement uses multiple VALUES clauses to add rows to NEWCOUNTRIES. Recall that NEWCOUNTRIES has six columns, so it is necessary to specify a value or an appropriate missing value for all six columns. See the results of the DESCRIBE TABLE statement in "Creating Tables like an Existing Table" on page 113 for information about the columns of NEWCOUNTRIES.

```
libname sql 'SAS-library';

proc sql;
   insert into sql.newcountries
      values ('Pakistan', 'Islamabad', 123060000, ., ' ', .)
      values ('Nigeria', 'Lagos', 99062000, ., ' ', .);
   title "World's Largest Countries";
   select name format=$20.,
          capital format=$15.,
          population format=comma15.0
      from sql.newcountries;
```

Output 4.3 *Rows Inserted with the Values Clause*

World's Largest Countries

Name	Capital	Population
Brazil	Brasilia	160,310,357
China	Beijing	1,202,215,077
India	New Delhi	929,009,120
Indonesia	Jakarta	202,393,859
Russia	Moscow	151,089,979
United States	Washington	263,294,808
Bangladesh	Dhaka	126,391,060
Japan	Tokyo	126,352,003
Pakistan	Islamabad	123,060,000
Nigeria	Lagos	99,062,000

Note the following features of VALUES clauses:

- As with other SQL clauses, use commas to separate columns. In addition, you must use a semicolon after the last VALUES clause only.
- If you omit data for a column without indicating a missing value, then you receive an error message and the row is not inserted.
- To specify that a value is missing, use a space in single quotation marks for character values and a period for numeric values.

Inserting Rows with a Query

You can insert the rows from a query result into a table. The following query returns rows for large countries (over 130 million in population) from the COUNTRIES table. The INSERT statement adds the data to the empty table NEWCOUNTRIES, which was created earlier in “Creating Tables like an Existing Table” on page 113:

```
libname sql 'SAS-library';

proc sql;
   create table sql.newcountries
      like sql.countries;

proc sql;
   title "World's Largest Countries";
   insert into sql.newcountries
   select * from sql.countries
      where population ge 130000000;

   select name format=$20.,
          capital format=$15.,
```

```
      population format=comma15.0
   from sql.newcountries;
```

Output 4.4 *Rows Inserted with a Query*

World's Largest Countries

Name	Capital	Population
Brazil	Brasilia	160,310,357
China	Beijing	1,202,215,077
India	New Delhi	929,009,120
Indonesia	Jakarta	202,393,859
Russia	Moscow	151,089,979
United States	Washington	263,294,808

If your query does not return data for every column, then you receive an error message, and the row is not inserted. For more information about how PROC SQL handles errors during data insertions, see "Handling Update Errors" on page 120.

Updating Data Values in a Table

You can use the UPDATE statement to modify data values in tables and in the tables that underlie PROC SQL and SAS/ACCESS views. For more information about updating views, see "Updating a View" on page 130. The UPDATE statement updates data in existing columns; it does not create new columns. To add new columns, see "Altering Columns" on page 121 and "Creating New Columns" on page 27. The examples in this section update the original NEWCOUNTRIES table.

Updating All Rows in a Column with the Same Expression

The following UPDATE statement increases all populations in the NEWCOUNTRIES table by 5%:

```
/* code for all examples in updating section */
libname sql 'SAS-library';

proc sql;
   delete from sql.newcountries;
   insert into sql.newcountries
   select * from sql.countries
      where population ge 130000000;


proc sql;
   update sql.newcountries
      set population=population*1.05;
   title "Updated Population Values";
   select name format=$20.,
          capital format=$15.,
          population format=comma15.0
      from sql.newcountries;
```

Output 4.5 *Updating a Column for All Rows*

Updated Population Values

Name	Capital	Population
Brazil	Brasilia	168,325,875
China	Beijing	1,262,325,831
India	New Delhi	975,459,576
Indonesia	Jakarta	212,513,552
Russia	Moscow	158,644,478
United States	Washington	276,459,548

Updating Rows in a Column with Different Expressions

If you want to update some, but not all, of a column's values, then use a WHERE expression in the UPDATE statement. You can use multiple UPDATE statements, each with a different expression. However, each UPDATE statement can have only one WHERE clause. The following UPDATE statements result in different population increases for different countries in the NEWCOUNRTRIES table.

```
libname sql 'SAS-library';

proc sql;
   delete from sql.newcountries;
   insert into sql.newcountries
   select * from sql.countries
      where population ge 130000000;

proc sql;
   update sql.newcountries
      set population=population*1.05
         where name like 'B%';

   update sql.newcountries
      set population=population*1.07
         where name in ('China', 'Russia');

   title "Selectively Updated Population Values";
   select name format=$20.,
          capital format=$15.,
          population format=comma15.0
      from sql.newcountries;
```

Output 4.6 *Selectively Updating a Column*

Selectively Updated Population Values

Name	Capital	Population
Brazil	Brasilia	168,325,875
China	Beijing	1,286,370,132
India	New Delhi	929,009,120
Indonesia	Jakarta	202,393,859
Russia	Moscow	161,666,278
United States	Washington	263,294,808

You can accomplish the same result with a CASE expression:

```
update sql.newcountries
   set population=population*
      case when name like 'B%' then 1.05
           when name in ('China', 'Russia') then 1.07
```

```
         else 1
      end;
```

If the WHEN clause is true, then the corresponding THEN clause returns a value that the SET clause then uses to complete its expression. In this example, when Name starts with the letter *B*, the SET expression becomes **population=population*1.05**.

CAUTION:

Make sure that you specify the ELSE clause. If you omit the ELSE clause, then each row that is not described in one of the WHEN clauses receives a missing value for the column that you are updating. This happens because the CASE expression supplies a missing value to the SET clause, and the Population column is multiplied by a missing value, which produces a missing value.

Handling Update Errors

While you are updating or inserting rows in a table, you might receive an error message that the update or insert cannot be performed. By using the UNDO_POLICY= option, you can control whether the changes that have already been made will be permanent.

The UNDO _POLICY= option in the PROC SQL and RESET statements determines how PROC SQL handles the rows that have been inserted or updated by the current INSERT or UPDATE statement up to the point of error.

UNDO_POLICY=REQUIRED
: is the default. It undoes all updates or inserts up to the point of error.

UNDO_POLICY=NONE
: does not undo any updates or inserts.

UNDO_POLICY=OPTIONAL
: undoes any updates or inserts that it can undo reliably.

Note: Alternatively, you can set the SQLUNDOPOLICY system option. For more information, see “SQLUNDOPOLICY= System Option” on page 368.

Deleting Rows

The DELETE statement deletes one or more rows in a table or in a table that underlies a PROC SQL or SAS/ACCESS view. For more information about deleting rows from views, see “Updating a View” on page 130. The following DELETE statement deletes the names of countries that begin with the letter *R*:

```
proc sql;
   delete from sql.newcountries;

   insert into sql.newcountries
   select * from sql.countries
      where population ge 130000000;

proc sql;
   delete
      from sql.newcountries
      where name like 'R%';
```

A note in the SAS log tells you how many rows were deleted.

Log 4.4 *SAS Log for DELETE Statement*

```
NOTE: 1 row was deleted from SQL.NEWCOUNTRIES.
```

Note: For PROC SQL tables, SAS deletes the data in the rows but retains the space in the table.

CAUTION:

If you omit a WHERE clause, then the DELETE statement deletes all the rows from the specified table or the table that is described by a view. The rows are not deleted from the table until it is recreated.

Altering Columns

The ALTER TABLE statement adds, modifies, and deletes columns in existing tables. You can use the ALTER TABLE statement with tables only; it does not work with views. A note appears in the SAS log that describes how you have modified the table.

Adding a Column

The ADD clause adds a new column to an existing table. You must specify the column name and data type. You can also specify a length (LENGTH=), format (FORMAT=), informat (INFORMAT=), and a label (LABEL=). The following ALTER TABLE statement adds the numeric data column Density to the NEWCOUNTRIES table:

```
proc sql;
   delete from sql.newcountries;
   insert into sql.newcountries
   select * from sql.countries
      where population ge 130000000;

proc sql;
   alter table sql.newcountries
      add density num label='Population Density' format=6.2;

   title "Population Density Table";
   select name format=$20.,
          capital format=$15.,
          population format=comma15.0,
          density
      from sql.newcountries;
```

Output 4.7 *Adding a New Column*

Population Density Table

Name	Capital	Population	Population Density
Brazil	Brasilia	160,310,357	.
China	Beijing	1,202,215,077	.
India	New Delhi	929,009,120	.
Indonesia	Jakarta	202,393,859	.
Russia	Moscow	151,089,979	.
United States	Washington	263,294,808	.

The new column is added to NEWCOUNTRIES, but it has no data values. The following UPDATE statement changes the missing values for Density from missing to the appropriate population densities for each country:

```
proc sql;
   update sql.newcountries
      set density=population/area;

   title "Population Density Table";
   select name format=$20.,
          capital format=$15.,
          population format=comma15.0,
          density
      from sql.newcountries;
```

Output 4.8 *Filling in the New Column's Values*

Population Density Table

Name	Capital	Population	Population Density
Brazil	Brasilia	160,310,357	48.78
China	Beijing	1,202,215,077	325.27
India	New Delhi	929,009,120	759.86
Indonesia	Jakarta	202,393,859	273.10
Russia	Moscow	151,089,979	22.92
United States	Washington	263,294,808	69.52

For more information about how to change data values, see "Updating Data Values in a Table" on page 118.

You can accomplish the same update by using an arithmetic expression to create the Population Density column as you recreate the table:

```
proc sql;
   create table sql.newcountries as
   select *, population/area as density
             label='Population Density'
             format=6.2
      from sql.newcountries;
```

See "Calculating Values" on page 29 for another example of creating columns with arithmetic expressions.

Modifying a Column

You can use the MODIFY clause to change the width, informat, format, and label of a column. To change a column's name, use the RENAME= data set option. You cannot change a column's data type by using the MODIFY clause.

The following MODIFY clause permanently changes the format for the Population column:

```
proc sql;
   delete from sql.newcountries;
   create table sql.newcountries as
   select * from sql.countries
      where population ge 130000000;


proc sql;
   title "World's Largest Countries";
   alter table sql.newcountries
      modify population format=comma15.;
   select name, population from sql.newcountries;
```

Output 4.9 *Modifying a Column Format*

World's Largest Countries

Name	Population
Brazil	160,310,357
China	1,202,215,077
India	929,009,120
Indonesia	202,393,859
Russia	151,089,979
United States	263,294,808

You might have to change a column's width (and format) before you can update the column. For example, before you can prefix a long text string to Name, you must change the width and format of Name from 35 to 60. The following statements modify and update the Name column:

```
proc sql;
   title "World's Largest Countries";
   alter table sql.newcountries
      modify name char(60) format=$60.;
   update sql.newcountries
      set name='The United Nations member country is '||name;

   select name from sql.newcountries;
```

Output 4.10 *Changing a Column's Width*

World's Largest Countries

Name
The United Nations member country is Brazil
The United Nations member country is China
The United Nations member country is India
The United Nations member country is Indonesia
The United Nations member country is Russia
The United Nations member country is United States

Deleting a Column

The DROP clause deletes columns from tables. The following DROP clause deletes UNDate from NEWCOUNTRIES:

```
proc sql;
   alter table sql.newcountries
      drop undate;
```

Creating an Index

An index is a file that is associated with a table. The index enables access to rows by index value. Indexes can provide quick access to small subsets of data, and they can enhance table joins. You can create indexes, but you cannot instruct PROC SQL to use an index. PROC SQL determines whether it is efficient to use the index. Some columns might not be appropriate for an index. In general, create indexes for columns that have many unique values or are columns that you use regularly in joins.

Using PROC SQL to Create Indexes

You can create a simple index, which applies to one column only. The name of a simple index must be the same as the name of the column that it indexes. Specify the column name in parentheses after the table name. The following CREATE INDEX statement creates an index for the Area column in NEWCOUNTRIES:

```
proc sql;
   create index area
      on sql.newcountries(area);
```

You can also create a composite index, which applies to two or more columns. The following CREATE INDEX statement creates the index Places for the Name and Continent columns in NEWCOUNTRIES:

```
proc sql;
   create index places
      on sql.newcountries(name, continent);
```

To ensure that each value of the indexed column (or each combination of values of the columns in a composite index) is unique, use the UNIQUE keyword:

```
proc sql;
   create unique index places
      on sql.newcountries(name, continent);
```

Using the UNIQUE keyword causes SAS to reject any change to a table that would cause more than one row to have the same index value.

Tips for Creating Indexes

- The name of the composite index cannot be the same as the name of one of the columns in the table.
- If you use two columns to access data regularly, such as a first name column and a last name column from an employee database, then you should create a composite index for the columns.
- Keep the number of indexes to a minimum to reduce disk space and update costs.
- Use indexes for queries that retrieve a relatively small number of rows (less than 15%).
- In general, indexing a small table does not result in a performance gain.
- In general, indexing on a column with a small number (less than 6 or 7) of distinct values does not result in a performance gain.
- You can use the same column in a simple index and in a composite index. However, for tables that have a primary key integrity constraint, do not create more than one index that is based on the same column as the primary key.

Deleting Indexes

To delete an index from a table, use the DROP INDEX statement. The following DROP INDEX statement deletes the index Places from NEWCOUNTRIES:

```
proc sql;
   drop index places from sql.newcountries;
```

Deleting a Table

To delete a PROC SQL table, use the DROP TABLE statement:

```
proc sql;
   drop table sql.newcountries;
```

Using SQL Procedure Tables in SAS Software

Because PROC SQL tables are SAS data files, you can use them as input to a DATA step or to other SAS procedures. For example, the following PROC MEANS step calculates the mean for Area for all countries in COUNTRIES:

```
proc means data=sql.countries mean maxdec=2;
   title "Mean Area for All Countries";
   var area;
run;
```

Output 4.11 *Using a PROC SQL Table in PROC MEANS*

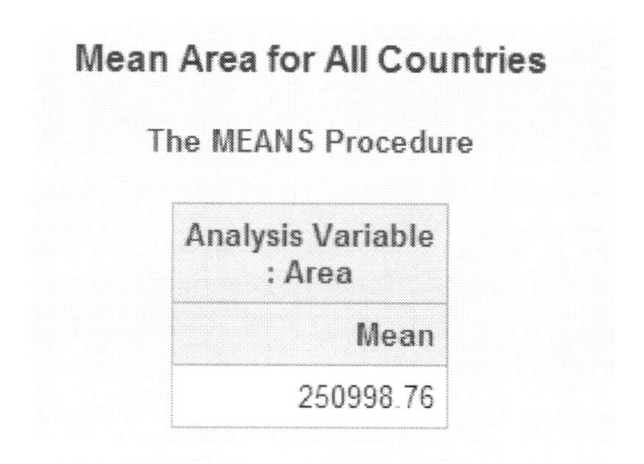

Mean Area for All Countries

The MEANS Procedure

Analysis Variable : Area
Mean
250998.76

Creating and Using Integrity Constraints in a Table

Integrity constraints are rules that you specify to guarantee the accuracy, completeness, or consistency of data in tables. All integrity constraints are enforced when you insert, delete, or alter data values in the columns of a table for which integrity constraints have been defined. Before a constraint is added to a table that contains existing data, all the data is checked to determine that it satisfies the constraints.

You can use general integrity constraints to verify that data in a column is one of the following:

- nonmissing
- unique

- both nonmissing and unique
- within a specified set or range of values

You can also apply referential integrity constraints to link the values in a specified column (called a **primary key**) of one table to values of a specified column in another table. When linked to a primary key, a column in the second table is called a **foreign key**.

When you define referential constraints, you can also choose what action occurs when a value in the primary key is updated or deleted.

- You can prevent the primary key value from being updated or deleted when matching values exist in the foreign key. This is the default.
- You can allow updates and deletions to the primary key values. By default, any affected foreign key values are changed to missing values. However, you can specify the CASCADE option to update foreign key values instead. Currently, the CASCADE option does not apply to deletions.

You can choose separate actions for updates and for deletions.

Note: Integrity constraints cannot be defined for views.

The following example creates integrity constraints for a table, MYSTATES, and another table, USPOSTAL. The constraints are as follows:

- state name must be unique and nonmissing in both tables
- population must be greater than 0
- continent must be either North America or Oceania

```
proc sql;
   create table sql.mystates
      (state      char(15),
       population num,
       continent  char(15),

          /* contraint specifications */
       constraint prim_key    primary key(state),
       constraint population  check(population gt 0),
       constraint continent   check(continent in ('North America', 'Oceania')));

   create table sql.uspostal
      (name      char(15),
       code      char(2) not null,          /* constraint specified as    */
                                            /* a column attribute         */

       constraint for_key foreign key(name) /* links NAME to the          */
                  references sql.mystates   /* primary key in MYSTATES    */

                     on delete restrict     /* forbids deletions to STATE */
                                            /* unless there is no         */
                                            /* matching NAME value        */

                     on update set null);   /* allows updates to STATE,   */
                                            /* changes matching NAME      */
                                            /* values to missing          */
```

The DESCRIBE TABLE statement displays the integrity constraints in the SAS log as part of the table description. The DESCRIBE TABLE CONSTRAINTS statement writes only the constraint specifications to the SAS log.

```
proc sql;
   describe table sql.mystates;
   describe table constraints sql.uspostal;
```

***Log 4.5** SAS Log Showing Integrity Constraints*

```
NOTE: SQL table SQL.MYSTATES was created like:

create table SQL.MYSTATES( bufsize=8192 )
  (
   state char(15),
   population num,
   continent char(15)
  );
create unique index state on SQL.MYSTATES(state);



               -----Alphabetic List of Integrity Constraints-----

  Integrity                        Where                      On        On
# Constraint Type         Variables Clause          Reference Delete    Update
-------------------------------------------------------------------------------
-49 continent   Check                 continent in

                                   ('North
                                   America',
                                   'Oceania')
-48 population Check                  population>0

-47 prim_key    Primary Key state

  for_key     Referential name                      SQL.      Restrict Set Null
                                                    USPOSTAL
NOTE: SQL table SQL.USPOSTAL ( bufsize=8192 ) has the following integrity
      constraints:



               -----Alphabetic List of Integrity Constraints-----

    Integrity                                                On          On
 #  Constraint   Type          Variables   Reference      Delete      Update
-----------------------------------------------------------------------------
 1  _NM0001_     Not Null      code
 2  for_key      Foreign Key   name        SQL.MYSTATES   Restrict    Set
Null
```

Integrity constraints cannot be used in views. For more information about integrity constraints, see *SAS Language Reference: Concepts*.

Creating and Using PROC SQL Views

Overview of Creating and Using PROC SQL Views

A PROC SQL view contains a stored query that is executed when you use the view in a SAS procedure or DATA step. Views are useful for the following reasons:

- often save space, because a view is frequently quite small compared with the data that it accesses
- prevent users from continually submitting queries to omit unwanted columns or row
- shield sensitive or confidential columns from users while enabling the same users to view other columns in the same table
- ensure that input data sets are always current, because data is derived from tables at execution time
- hide complex joins or queries from users

Creating Views

To create a PROC SQL view, use the CREATE VIEW statement, as shown in the following example:

```
libname sql 'SAS-library';

proc sql;
   title 'Current Population Information for Continents';
   create view sql.newcontinents as
   select continent,
          sum(population) as totpop  format=comma15. label='Total Population',
          sum(area) as totarea format=comma15. label='Total Area'
     from sql.countries
     group by continent;

   select * from sql.newcontinents;
```

Output 4.12 *An SQL Procedure View*

Current Population Information for Continents

Continent	Total Population	Total Area
	384,772	876,800
Africa	710,529,592	11,299,595
Asia	3,381,858,879	12,198,325
Australia	18,255,944	2,966,200
Central America and Caribbean	66,815,930	291,463
Europe	813,481,724	9,167,250
North America	384,801,818	8,393,092
Oceania	5,342,368	129,600
South America	317,568,801	6,885,418

Note: In this example, each column has a name. If you are planning to use a view in a procedure that requires variable names, then you must supply column aliases that you can reference as variable names in other procedures. For more information, see "Using SQL Procedure Views in SAS Software" on page 134.

Describing a View

The DESCRIBE VIEW statement writes a description of the PROC SQL view to the SAS log. The following SAS log describes the view NEWCONTINENTS, which is created in "Creating Views" on page 129:

```
proc sql;
   describe view sql.newcontinents;
```

Log 4.6 *SAS Log from DESCRIBE VIEW Statement*

```
NOTE: SQL view SQL.NEWCONTINENTS is defined as:

        select continent, SUM(population) as totpop label='Total Population'
format=COMMA15.0, SUM(area) as totarea label='Total Area' format=COMMA15.0
          from SQL.COUNTRIES
      group by continent;
```

Updating a View

You can update data through a PROC SQL and SAS/ACCESS view with the INSERT, DELETE, and UPDATE statements, under the following conditions.

- You can update only a single table through a view. The underlying table cannot be joined to another table or linked to another table with a set operator. The view cannot contain a subquery.

- If the view accesses a DBMS table, then you must have been granted the appropriate authorization by the external database management system (for example, ORACLE). You must have installed the SAS/ACCESS software for your DBMS. See the SAS/ACCESS documentation for your DBMS for more information about SAS/ACCESS views.
- You can update a column in a view by using the column's alias, but you cannot update a derived column, that is, a column that is produced by an expression. In the following example, you can update SquareMiles, but not Density:

```
proc sql;
   create view mycountries as
      select Name,
             area as SquareMiles,
             population/area as Density
         from sql.countries;
```

- You can update a view that contains a WHERE clause. The WHERE clause can be in the UPDATE clause or in the view. You cannot update a view that contains any other clause, such as ORDER BY, HAVING, and so on.

Embedding a LIBNAME in a View

You can embed a SAS LIBNAME statement or a SAS/ACCESS LIBNAME statement in a view by using the USING LIBNAME clause. When PROC SQL executes the view, the stored query assigns the libref. For SAS/ACCESS librefs, PROC SQL establishes a connection to a DBMS. The scope of the libref is local to the view and does not conflict with any identically named librefs in the SAS session. When the query finishes, the libref is disassociated. The connection to the DBMS is terminated and all data in the library becomes unavailable.

The advantage of embedded librefs is that you can store engine-host options and DBMS connection information, such as passwords, in the view. That, in turn, means that you do not have to remember and reenter that information when you want to use the libref.

Note: The USING LIBNAME clause must be the last clause in the SELECT statement. Multiple clauses can be specified, separated by commas.

In the following example, the libref OILINFO is assigned and a connection is made to an ORACLE database:

```
proc sql;
   create view sql.view1 as
      select *
         from oilinfo.reserves as newreserves
         using libname oilinfo oracle
            user=username
            pass=password
            path='dbms-path';
```

For more information about the SAS/ACCESS LIBNAME statement, see the SAS/ACCESS documentation for your DBMS.

The following example embeds a SAS LIBNAME statement in a view:

```
proc sql;
   create view sql.view2 as
      select *
         from oil.reserves
         using libname oil 'SAS-data-library';
```

Deleting a View

To delete a view, use the DROP VIEW statement:

```
proc sql;
   drop view sql.newcontinents;
```

Specifying In-Line Views

In some cases, you might want to use a query in a FROM clause instead of a table or view. You could create a view and refer to it in your FROM clause, but that process involves two steps. To save the extra step, specify the view in-line, enclosed in parentheses, in the FROM clause.

An in-line view is a query that appears in the FROM clause. An in-line view produces a table internally that the outer query uses to select data. Unlike views that are created with the CREATE VIEW statement, in-line views are not assigned names and cannot be referenced in other queries or SAS procedures as if they were tables. An in-line view can be referenced only in the query in which it is defined.

In the following query, the populations of all Caribbean and Central American countries are summed in an in-line query. The WHERE clause compares the sum with the populations of individual countries. Only countries that have a population greater than the sum of Caribbean and Central American populations are displayed.

```
libname sql 'SAS-library';

proc sql;
   title 'Countries With Population GT Caribbean Countries';
   select w.Name, w.Population format=comma15., c.TotCarib
      from (select sum(population) as TotCarib format=comma15.
                 from sql.countries
                where continent = 'Central America and Caribbean') as c,
           sql.countries as w
        where w.population gt c.TotCarib;
```

Output 4.13 *Using an In-Line View*

Countries With Population GT Caribbean Countries

Name	Population	TotCarib
Bangladesh	126,387,850	66,815,930
Brazil	160,310,357	66,815,930
China	1,202,215,077	66,815,930
Germany	81,890,690	66,815,930
India	929,009,120	66,815,930
Indonesia	202,393,859	66,815,930
Japan	126,345,434	66,815,930
Mexico	93,114,708	66,815,930
Nigeria	99,062,003	66,815,930
Pakistan	123,062,252	66,815,930
Philippines	70,500,039	66,815,930
Russia	151,089,979	66,815,930
United States	263,294,808	66,815,930
Vietnam	73,827,657	66,815,930

Tips for Using SQL Procedure Views

- Avoid using an ORDER BY clause in a view. If you specify an ORDER BY clause, then the data must be sorted each time that the view is referenced.
- If data is used many times in one program or in multiple programs, then it is more efficient to create a table rather than a view. If a view is referenced often in one program, then the data must be accessed at each reference.
- If the view resides in the same SAS library as the contributing table or tables, then specify a one-level name in the FROM clause. The default for the libref for the FROM clause's table or tables is the libref of the library that contains the view. This prevents you from having to change the view if you assign a different libref to the SAS library that contains the view and its contributing table or tables. This tip is used in the view that is described in "Creating Views" on page 129.
- Avoid creating views that are based on tables whose structure might change. A view is no longer valid when it references a nonexistent column.
- When you process PROC SQL views between a client and a server, getting the correct results depends on the compatibility between the client and server architecture. For more information, see "Accessing a SAS View" in Chapter 17 of *SAS/CONNECT User's Guide*.

Using SQL Procedure Views in SAS Software

You can use PROC SQL views as input to a DATA step or to other SAS procedures. The syntax for using a PROC SQL view in SAS is the same as that for a PROC SQL table. For an example, see "Using SQL Procedure Tables in SAS Software" on page 126.

Chapter 5

Programming with the SQL Procedure

Introduction

This section shows you how to do the following:

- use PROC SQL options to create and debug queries
- improve query performance
- access dictionary tables and how they are useful in gathering information about the elements of SAS
- use PROC SQL with the SAS macro facility
- use PROC SQL with the REPORT procedure
- access DBMSs by using SAS/ACCESS software
- format PROC SQL output by using the SAS Output Delivery System (ODS)

Using PROC SQL Options to Create and Debug Queries

Overview of Using PROC SQL Options to Create and Debug Queries

PROC SQL supports options that can give you greater control over PROC SQL while you are developing a query:

- The INOBS=, OUTOBS=, and LOOPS= options reduce query execution time by limiting the number of rows and the number of iterations that PROC SQL processes.
- The EXEC and VALIDATE statements enable you to quickly check the syntax of a query.
- The FEEDBACK option expands a SELECT * statement into a list of columns that the statement represents.
- The PROC SQL STIMER option records and displays query execution time.

You can set an option initially in the PROC SQL statement, and then use the RESET statement to change the same option's setting without ending the current PROC SQL step.

Restricting Row Processing with the INOBS= and OUTOBS= Options

When you are developing queries against large tables, you can reduce the time that it takes for the queries to run by reducing the number of rows that PROC SQL processes. Subsetting the tables with WHERE statements is one way to do this. Using the INOBS= and the OUTOBS= options are other ways.

The INOBS= option restricts the number of rows that PROC SQL takes as input from any single source. For example, if you specify INOBS=10, then PROC SQL uses only 10 rows from any table or view that is specified in a FROM clause. If you specify INOBS=10 and join two tables without using a WHERE clause, then the resulting table (Cartesian product) contains a maximum of 100 rows. The INOBS= option is similar to the SAS system option OBS=.

The OUTOBS= option restricts the number of rows that PROC SQL displays or writes to a table. For example, if you specify OUTOBS=10 and insert values into a table by using a query, then PROC SQL inserts a maximum of 10 rows into the resulting table. OUTOBS= is similar to the SAS data set option OBS=.

In a simple query, there might be no apparent difference between using INOBS or OUTOBS. However, at other times it is important to choose the correct option. For example, taking the average of a column with INOBS=10 returns an average of only 10 values from that column.

Limiting Iterations with the LOOPS= Option

The LOOPS= option restricts PROC SQL to the number of iterations that are specified in this option through its inner loop. By setting a limit, you can prevent queries from consuming excessive computer resources. For example, joining three large tables without meeting the join-matching conditions could create a huge internal table that would be inefficient to process. Use the LOOPS= option to prevent this from happening.

You can use the number of iterations that are reported in the SQLOOPS macro variable (after each PROC SQL statement is executed) to gauge an appropriate value for the LOOPS= option. For more information, see “Using the PROC SQL Automatic Macro Variables” on page 157.

If you use the PROMPT option with the INOBS=, OUTOBS=, or LOOPS= options, you are prompted to stop or continue processing when the limits set by these options are reached.

Checking Syntax with the NOEXEC Option and the VALIDATE Statement

To check the syntax of a PROC SQL step without actually executing it, use the NOEXEC option or the VALIDATE statement. The NOEXEC option can be used once in the PROC SQL statement, and the syntax of all queries in that PROC SQL step will be checked for accuracy without executing them. The VALIDATE statement must be specified before each SELECT statement in order for that statement to be checked for accuracy without executing. If the syntax is valid, then a message is written to the SAS log to that effect. If the syntax is invalid, then an error message is displayed. The automatic macro variable SQLRC contains an error code that indicates the validity of the syntax. For an example of the VALIDATE statement used in PROC SQL, see “Validating a Query” on page 71. For an example of using the VALIDATE statement in a SAS/AF application, see “Using the PROC SQL Automatic Macro Variables” on page 157.

Note: There is an interaction between the PROC SQL EXEC and ERRORSTOP options when SAS is running in a batch or noninteractive session. For more information, see Chapter 7, “SQL Procedure,” on page 209.

Expanding SELECT * with the FEEDBACK Option

The FEEDBACK option expands a SELECT * (ALL) statement into the list of columns that the statement represents. Any PROC SQL view is expanded into the underlying query, all expressions are enclosed in parentheses to indicate their order of evaluation, and the PUT function optimizations that are performed on the query are displayed. The FEEDBACK option also displays the resolved values of macros and macro variables.

For example, the following query is expanded in the SAS log:

```
libname sql 'SAS-library';

proc sql feedback;
   select * from sql.countries;
```

Log 5.1 *Expanded SELECT * Statement*

```
NOTE: Statement transforms to:

        select COUNTRIES.Name, COUNTRIES.Capital, COUNTRIES.Population,
COUNTRIES.Area, COUNTRIES.Continent, COUNTRIES.UNDate
          from SQL.COUNTRIES;
```

Timing PROC SQL with the STIMER Option

Certain operations can be accomplished in more than one way. For example, there is often a join equivalent to a subquery. Consider factors such as readability and maintenance, but generally you will choose the query that runs fastest. The SAS system option STIMER shows you the cumulative time for an entire procedure. The PROC SQL STIMER option shows you how fast the individual statements in a PROC SQL step are running. This enables you to optimize your query.

Note: For the PROC SQL STIMER option to work, the SAS system option STIMER must also be specified.

This example compares the execution times of two queries. Both queries list the names and populations of states in the UNITEDSTATES table that have a larger population than Belgium. The first query does this with a join; the second with a subquery. Log 5.2 on page 139 shows the STIMER results from the SAS log.

```
libname sql 'SAS-library';

proc sql stimer ;
   select us.name, us.population
      from sql.unitedstates as us, sql.countries as w
      where us.population gt w.population and
            w.name = 'Belgium';

   select Name, population
      from sql.unitedstates
      where population gt
                (select population from sql.countries
                   where name = 'Belgium');
```

Log 5.2 *Comparing Run Times of Two Queries*

```
4  proc sql stimer ;
NOTE: SQL Statement used:
      real time           0.00 seconds
      cpu time            0.01 seconds

5     select us.name, us.population
6        from sql.unitedstates as us, sql.countries as w
7        where us.population gt w.population and
8              w.name = 'Belgium';
NOTE: The execution of this query involves performing one or more Cartesian
      product joins that can not be optimized.
NOTE: SQL Statement used:
      real time           0.10 seconds
      cpu time            0.05 seconds

9
10     select Name, population
11        from sql.unitedstates
12        where population gt
13                (select population from sql.countries
14                    where name = 'Belgium');
NOTE: SQL Statement used:
      real time           0.09 seconds
      cpu time            0.09 seconds
```

Compare the CPU time of the first query (that uses a join), 0.05 seconds, with 0.09 seconds for the second query (that uses a subquery). Although there are many factors that influence the run times of queries, generally a join runs faster than an equivalent subquery.

Resetting PROC SQL Options with the RESET Statement

Use the RESET statement to add, drop, or change the options in the PROC SQL statement. You can list the options in any order in the PROC SQL and RESET statements. Options stay in effect until they are reset.

This example first uses the NOPRINT option to prevent the SELECT statement from displaying its result table in SAS output. The RESET statement then changes the NOPRINT option to PRINT (the default) and adds the NUMBER option, which displays the row number in the result table.

```
proc sql noprint;
   title 'Countries with Population Under 20,000';
   select Name, Population from sql.countries;
reset print number;
   select Name, Population from sql.countries
      where population lt 20000;
```

Output 5.1 *Resetting PROC SQL Options with the RESET Statement*

Countries with Population Under 20,000

Row	Name	Population
1	Leeward Islands	12119
2	Nauru	10099
3	Turks and Caicos Islands	12119
4	Tuvalu	10099
5	Vatican City	1010

Improving Query Performance

Overview of Improving Query Performance

There are several ways to improve query performance, including the following:

- using indexes and composite indexes
- using the keyword ALL in set operations when you know that there are no duplicate rows, or when it does not matter if you have duplicate rows in the result table
- omitting the ORDER BY clause when you create tables and views
- using in-line views instead of temporary tables (or vice versa)
- using joins instead of subqueries
- using WHERE expressions to limit the size of result tables that are created with joins
- using either PROC SQL options, SAS system options, or both to replace a PUT function in a query with a logically equivalent expression
- replacing references to the DATE, TIME, DATETIME, and TODAY functions in a query with their equivalent constant values before the query executes
- disabling the remerging of data when summary functions are used in a query

Using Indexes to Improve Performance

Indexes are created with the CREATE INDEX statement in PROC SQL or with the MODIFY and INDEX CREATE statements in the DATASETS procedure. Indexes are stored in specialized members of a SAS library and have a SAS member type of INDEX. The values that are stored in an index are automatically updated if you make a change to the underlying data.

Indexes can improve the performance of certain classes of retrievals. For example, if an indexed column is compared to a constant value in a WHERE expression, then the index will likely improve the query's performance. Indexing the column that is specified in a correlated reference to an outer table also improves a subquery's (and hence, query's) performance. Composite indexes can improve the performance of queries that compare the columns that are named in the composite index with constant values that are linked using the AND operator. For example, if you have a compound index in the columns CITY and STATE, and the WHERE expression is specified as WHERE CITY='xxx' AND STATE='yy', then the index can be used to select that subset of rows more efficiently. Indexes can also benefit queries that have a WHERE clause in this form:

```
... where var1 in (select item1 from table1) ...
```

The values of VAR1 from the outer query are found in the inner query by using the index. An index can improve the processing of a table join, if the columns that participate in the join are indexed in one of the tables. This optimization can be done for equijoin queries only—that is, when the WHERE expression specifies that table1.X=table2.Y.

Using the Keyword ALL in Set Operations

Set operators such as UNION, OUTER UNION, EXCEPT, and INTERSECT can be used to combine queries. Specifying the optional ALL keyword prevents the final process that eliminates duplicate rows from the result table. You should use the ALL form when you know that there are no duplicate rows or when it does not matter whether the duplicate rows remain in the result table.

Omitting the ORDER BY Clause When Creating Tables and Views

If you specify the ORDER BY clause when a table or view is created, then the data is always displayed in that order unless you specify another ORDER BY clause in a query that references that table or view. As with any sorting procedure, using ORDER BY when retrieving data has certain performance costs, especially on large tables. If the order of your output is not important for your results, then your queries will typically run faster without an ORDER BY clause.

Using In-Line Views versus Temporary Tables

It is often helpful when you are exploring a problem to break a query down into several steps and create temporary tables to hold the intermediate results. After you have worked through the problem, combining the queries into one query by using in-line views can be more efficient. However, under certain circumstances it is more efficient to use temporary tables. You should try both methods to determine which is more efficient for your case.

Comparing Subqueries with Joins

Many subqueries can also be expressed as joins. Generally, a join is processed at least as efficiently as the subquery. PROC SQL stores the result values for each unique set of correlation columns temporarily, thereby eliminating the need to calculate the subquery more than once.

Using WHERE Expressions with Joins

When joining tables, you should specify a WHERE expression. Joins without WHERE expressions are often time-consuming to evaluate because of the multiplier effect of the Cartesian product. For example, joining two tables of 1,000 rows each without specifying a WHERE expression or an ON clause, produces a result table with one million rows.

PROC SQL executes and obtains the correct results in unbalanced WHERE expressions (or ON join expressions) in an equijoin, as shown here, but handles them inefficiently:

```
where table1.columnA-table2.columnB=0
```

It is more efficient to rewrite this clause to balance the expression so that columns from each table are on alternate sides of the equals condition:

```
where table1.columnA=table2.columnB
```

PROC SQL sequentially processes joins that do not have an equijoin condition evaluating each row against the WHERE expression: that is, joins without an equijoin condition are not evaluated using sort-merge or index-lookup techniques. Evaluating left and right outer joins is generally comparable to, or only slightly slower than, a standard inner join. A full outer join usually requires two passes over both tables in the join, although PROC SQL tries to store as much data as possible in buffers. Thus for small tables, an outer join might be processed with only one physical read of the data.

Optimizing the PUT Function

Reducing the PUT Function

There are several ways that you can improve the performance of a query by optimizing the PUT function. If you reference tables in a database, eliminating references to PUT functions can enable more of the query to be passed to the database. It can simplify SELECT statement evaluation for the default Base SAS engine.

There are five possible evaluations that are performed when optimizing the PUT function:

- Functions, including PUT, that contain literal values.
- PUT functions in the WHERE and HAVING clauses that contain formats that are supplied by SAS.
- PUT functions in the WHERE and HAVING clauses that contain user-defined formats.
- PUT functions in any part of the SELECT statement that contain user-defined formats that are defined with an OTHER= clause.
- PUT functions that are deployed inside the database.

Controlling PUT Function Optimization

- If you specify either the PROC SQL REDUCEPUT= option or the SQLREDUCEPUT= system option, SAS optimizes the PUT function before the query is executed.

  The following SELECT statements are examples of queries that would be optimized:

  ```
  select x, y from sqllibb where (PUT(x, abc.) in ('yes', 'no'));
  select x from sqlliba where (PUT(x, udfmt.) = trim(left('small')));
  ```

- For databases that allow implicit pass-through when the row count for a table is not known, PROC SQL allows the optimization in order for the query to be executed by the database. When the PROC SQL REDUCEPUT= option or the SQLREDUCEPUT= system option is set to DBMS, BASE, or ALL, PROC SQL considers the value of the PROC SQL REDUCEPUTOBS= option or the SQLREDUCEPUTOBS= system option and determines whether to optimize the PUT function. The PROC SQL REDUCEPUTOBS= option or the SQLREDUCEPUTOBS= system option specifies the minimum number of rows that must be in a table in order for PROC SQL to consider optimizing the PUT function in a query. For databases that do not allow implicit pass-through, PROC SQL does not perform the optimization, and more of the query is performed by SAS.

- Some formats, especially user-defined formats, can contain many format values. Depending on the number of matches for a given PUT function expression, the resulting expression can list many format values. If the number of format values becomes too large, the query performance can degrade. When the PROC SQL REDUCEPUT= option or the SQLREDUCEPUT= system option is set to DBMS, BASE, or ALL, PROC SQL considers the value of the PROC SQL REDUCEPUTVALUES= option or the SQLREDUCEPUTVALUES= system option and determines whether to optimize the PUT function in a query. For databases that do not allow implicit pass-through, PROC SQL does not perform the optimization, and more of the query is performed by SAS.

For more information, see the REDUCEPUT=, REDUCEPUTOBS=, and REDUCEPUTVALUES= options in Chapter 7, "SQL Procedure," on page 209, and the SQLREDUCEPUT=, SQLREDUCEPUTOBS=, and SQLREDUCEPUTVALUES= system options in Appendix 1, "SQL Macro Variables and System Options," on page 359.

Note: PROC SQL can consider both the REDUCEPUTOBS= and the REDUCEPUTVALUES= options (or SQLREDUCEPUTOBS= and SQLREDUCEPUTVALUES= system options) when trying to determine whether to optimize the PUT function.

Deploying the PUT Function and SAS Formats inside a DBMS

SAS/ACCESS software for relational databases enables you to use the format publishing macro to deploy or publish the PUT function implementation to the database as a function named SAS_PUT(). As with any other programming function, the SAS_PUT() function can take one or more input parameters and return an output value. The default value for the SQLMAPPUTTO system option is SAS_PUT. After the SAS_PUT() function is deployed in the database, you can use the SAS_PUT() function as you would use any standard SQL function inside the database.

In addition, the SAS_PUT() function supports the use of SAS formats in SQL queries that are submitted to the database. You can use the format publishing macro to publish to the database both the formats that are supplied by SAS and the custom formats that you create with the FORMAT procedure.

By publishing the PUT function implementation to the database as the SAS_PUT() function to support the use of SAS formats, and by packaging both the formats that are supplied by SAS and the custom formats that you create with the FORMAT procedure, the following advantages are realized:

- The entire SQL query can be processed inside the database.
- The SAS format processing leverages the DBMS's scalable architecture.
- The results are grouped by the formatted data, and are extracted from the database.

Note: If you use the SQL_FUNCTIONS= LIBNAME statement option to remap the PUT function (for example, SAS_PUT( )), then the SQL_FUNCTIONS= LIBNAME option takes precedence over the SQLMAPPUTTO= system option. For more information, see "SQL_FUNCTIONS= LIBNAME Option" in *SAS/ACCESS for Relational Databases: Reference.*

TIP Using both the SQLREDUCEPUT= system option (or the PROC SQL REDUCEPUT= option) and the SAS_PUT() function can result in a significant performance boost.

For more information about using the In-database format publishing macro and the SQLMAPPUTTO system option, see *SAS/ACCESS for Relational Databases: Reference*.

Replacing References to the DATE, TIME, DATETIME, and TODAY Functions

When the PROC SQL CONSTDATETIME option or the SQLCONSTDATETIME system option is set, PROC SQL evaluates the DATE, TIME, DATETIME, and TODAY functions in a query once, and uses those values throughout the query. Computing these values once ensures consistent results when the functions are used multiple times in a query, or when the query executes the functions close to a date or time boundary. When referencing database tables, performance is enhanced because it allows more of the query to be passed down to the database.

For more information, see the "SQLCONSTDATETIME System Option" on page 359 or the CONSTDATETIME option in the *Base SAS Procedures Guide*.

Note: If you specify both the PROC SQL REDUCEPUT option or the SQLREDUCEPUT= system option and the PROC SQL CONSTDATETIME option or the SQLCONSTDATETIME system option, PROC SQL replaces the DATE, TIME, DATETIME, and TODAY functions with their respective values in order to determine the PUT function value before the query executes.

Disabling the Remerging of Data When Using Summary Functions

When you use a summary function in a SELECT clause or a HAVING clause, PROC SQL might remerge the data. Remerging the data involves two passes through the data. If you set the PROC SQL NOREMERGE option or the NOSQLREMERGE system option, PROC SQL will not process the remerging of data. When referencing database tables, performance is enhanced because it enables more of the query to be passed down to the database.

For more information, see the PROC SQL statement REMERGE option in the *Base SAS Procedures Guide* and the SQLREMERGE system option in Appendix 1, "SQL Macro Variables and System Options," on page 359.

Accessing SAS System Information by Using DICTIONARY Tables

What Are Dictionary Tables?

DICTIONARY tables are special read-only PROC SQL tables or views. They retrieve information about all the SAS libraries, SAS data sets, SAS system options, and external files that are associated with the current SAS session. For example, the DICTIONARY.COLUMNS table contains information such as name, type, length, and format, about all columns in all tables that are known to the current SAS session.

PROC SQL automatically assigns the DICTIONARY libref. To get information from DICTIONARY tables, specify DICTIONARY.table-name in the FROM clause in a SELECT statement in PROC SQL.

DICTIONARY.table-name is valid in PROC SQL only. However, SAS provides PROC SQL views, based on the DICTIONARY tables, that can be used in other SAS procedures and in the DATA step. These views are stored in the SASHELP library and are commonly called "SASHELP views."

For an example of a DICTIONARY table, see "Example 6: Reporting from DICTIONARY Tables" on page 257.

The following table describes the DICTIONARY tables that are available and shows the associated SASHELP views for each table.

Table 5.1 *DICTIONARY Tables and Associated SASHELP Views*

DICTIONARY Table	SASHELP View	Description
CATALOGS	VCATALG	Contains information about known SAS catalogs.
CHECK_CONSTRAINTS	VCHKCON	Contains information about known check constraints.
COLUMNS	VCOLUMN	Contains information about columns in all known tables.
CONSTRAINT_COLUMN_USAGE	VCNCOLU	Contains information about columns that are referred to by integrity constraints.
CONSTRAINT_TABLE_USAGE	VCNTABU	Contains information about tables that have integrity constraints defined on them.
DATAITEMS	VDATAIT	Contains information about known information map data items.
DESTINATIONS	VDEST	Contains information about known ODS destinations.
DICTIONARIES	VDCTNRY	Contains information about all DICTIONARY tables.
ENGINES	VENGINE	Contains information about SAS engines.
EXTFILES	VEXTFL	Contains information about known external files.
FILTERS	VFILTER	Contains information about known information map filters.
FORMATS	VFORMAT VCFORMAT	Contains information about currently accessible formats and informats.
FUNCTIONS	VFUNC	Contains information about currently accessible functions.
GOPTIONS	VGOPT VALLOPT	Contains information about currently defined graphics options (SAS/GRAPH software). SASHELP.VALLOPT includes SAS system options as well as graphics options.
INDEXES	VINDEX	Contains information about known indexes.
INFOMAPS	VINFOMP	Contains information about known information maps.

DICTIONARY Table	SASHELP View	Description
LIBNAMES	VLIBNAM	Contains information about currently defined SAS libraries.
MACROS	VMACRO	Contains information about currently defined macro variables.
MEMBERS	VMEMBER VSACCES VSCATLG VSLIB VSTABLE VSTABVW VSVIEW	Contains information about all objects that are in currently defined SAS libraries. SASHELP.VMEMBER contains information for all member types; the other SASHELP views are specific to particular member types (such as tables or views).
OPTIONS	VOPTION VALLOPT	Contains information about SAS system options. SASHELP.VALLOPT includes graphics options as well as SAS system options.
REFERENTIAL_CONSTRAINTS	VREFCON	Contains information about referential constraints.
REMEMBER	VREMEMB	Contains information about known remembers.
STYLES	VSTYLE	Contains information about known ODS styles.
TABLE_CONSTRAINTS	VTABCON	Contains information about integrity constraints in all known tables.
TABLES	VTABLE	Contains information about known tables.
TITLES	VTITLE	Contains information about currently defined titles and footnotes.
VIEWS	VVIEW	Contains information about known data views.
VIEW_SOURCES	Not available	Contains a list of tables (or other views) referenced by the SQL or DATASTEP view, and a count of the number of references.

Retrieving Information about DICTIONARY Tables and SASHELP Views

To see how each DICTIONARY table is defined, submit a DESCRIBE TABLE statement. This example shows the definition of DICTIONARY.TABLES:

```
proc sql;
   describe table dictionary.tables;
```

The results are written to the SAS log.

***Log 5.3** Definition of DICTIONARY.TABLES*

```
NOTE: SQL table DICTIONARY.TABLES was created like:

create table DICTIONARY.TABLES
  (
   libname char(8) label='Library Name',
   memname char(32) label='Member Name',
   memtype char(8) label='Member Type',
   dbms_memtype char(32) label='DBMS Member Type',
   memlabel char(256) label='Data Set Label',
   typemem char(8) label='Data Set Type',
   crdate num format=DATETIME informat=DATETIME label='Date Created',
   modate num format=DATETIME informat=DATETIME label='Date Modified',
   nobs num label='Number of Physical Observations',
   obslen num label='Observation Length',
   nvar num label='Number of Variables',
   protect char(3) label='Type of Password Protection',
   compress char(8) label='Compression Routine',
   encrypt char(8) label='Encryption',
   npage num label='Number of Pages',
   filesize num label='Size of File',
   pcompress num label='Percent Compression',
   reuse char(3) label='Reuse Space',
   bufsize num label='Bufsize',
   delobs num label='Number of Deleted Observations',
   nlobs num label='Number of Logical Observations',
   maxvar num label='Longest variable name',
   maxlabel num label='Longest label',
   maxgen num label='Maximum number of generations',
   gen num label='Generation number',
   attr char(3) label='Data Set Attributes',
   indxtype char(9) label='Type of Indexes',
   datarep char(32) label='Data Representation',
   sortname char(8) label='Name of Collating Sequence',
   sorttype char(4) label='Sorting Type',
   sortchar char(8) label='Charset Sorted By',
   reqvector char(24) format=$HEX48 informat=$HEX48 label='Requirements Vector',
   datarepname char(170) label='Data Representation Name',
   encoding char(256) label='Data Encoding',
   audit char(8) label='Audit Trail Active?',
   audit_before char(8) label='Audit Before Image?',
   audit_admin char(8) label='Audit Admin Image?',
   audit_error char(8) label='Audit Error Image?',
   audit_data char(8) label='Audit Data Image?',
   num_character num label='Number of Character Variables',
   num_numeric num label='Number of Numeric Variables'
  );
```

Similarly, you can use the DESCRIBE VIEW statement in PROC SQL to determine how a SASHELP view is defined. Here is an example:

```
proc sql;
   describe view sashelp.vtable;
```

Log 5.4 Description of SASHELP.VTABLE

```
NOTE: SQL view SASHELP.VSTABVW is defined as:

        select libname, memname, memtype
          from DICTIONARY.MEMBERS
         where (memtype='VIEW') or (memtype='DATA')
      order by libname asc, memname asc;
```

Using DICTIONARY.TABLES

DICTIONARY tables are commonly used to monitor and manage SAS sessions because the data is more easily manipulated than the output from other sources such as PROC DATASETS. You can query DICTIONARY tables the same way you query any other table, including subsetting with a WHERE clause, ordering the results, and creating PROC SQL views.

Note that many character values in the DICTIONARY tables are stored as all-uppercase characters; you should design your queries accordingly.

Because DICTIONARY tables are read-only objects, you cannot insert rows or columns, alter column attributes, or add integrity constraints to them.

Note: For DICTIONARY.TABLES and SASHELP.VTABLE, if a table is read-protected with a password, then the only information that is listed for that table is the library name, member name, member type, and type of password protection. All other information is set to missing.

Note: An error occurs if DICTIONARY.TABLES is used to retrieve information about an SQL view that exists in one library but has an input table from a second library that has not been assigned.

The following query uses a SELECT and subsetting WHERE clause to retrieve information about permanent tables and views that appear in the SQL library:

```
libname sql '\\sashq\root\pub\pubdoc\doc\901\authoring\sqlproc\miscsrc\sasfiles\';
options nodate nonumber linesize=80 pagesize=60;

   libname sql 'SAS-library';

   proc sql;
      title 'All Tables and Views in the SQL Library';
      select libname, memname, memtype, nobs
         from dictionary.tables
         where libname='SQL';
```

Output 5.2 *Tables and Views Used in This Document*

All Tables and Views in the SQL Library

Library Name	Member Name	Member Type	Number of Physical Observations
SQL	A	DATA	4
SQL	ACT	DATA	10
SQL	B	DATA	3
SQL	CAL	DATA	2
SQL	CONTINENTS	DATA	9
SQL	COUNTRIES	DATA	208
SQL	DENSITIES	DATA	10
SQL	FEATURES	DATA	74
SQL	HOL	DATA	2
SQL	MYSTATES	DATA	0
SQL	NEWCONTINENTS	VIEW	.
SQL	NEWCOUNTRIES	DATA	6
SQL	NEWSTATES	DATA	0
SQL	OILPROD	DATA	19
SQL	OILRSRVS	DATA	19
SQL	ONE	DATA	2
SQL	POSTALCODES	DATA	58
SQL	TWO	DATA	3
SQL	UNITEDSTATES	DATA	51
SQL	USCITYCOORDS	DATA	132
SQL	USPOSTAL	DATA	0
SQL	WOR	DATA	2
SQL	WORLDCITYCOORDS	DATA	212
SQL	WORLDTEMPS	DATA	59

Using DICTIONARY.COLUMNS

DICTIONARY tables are useful when you want to find specific columns to include in reports. The following query shows which of the tables that are used in this document contain the Country column:

```
libname sql 'SAS-library';
```

```
proc sql;
   title 'All Tables That Contain the Country Column';
   select libname, memname, name
      from dictionary.columns
      where name='Country' and
            libname='SQL';
```

Output 5.3 *Using DICTONARY.COLUMNS to Locate Specific Columns*

All Tables That Contain the Country Column

Library Name	Member Name	Column Name
SQL	OILPROD	Country
SQL	OILRSRVS	Country
SQL	WORLDCITYCOORDS	Country
SQL	WORLDTEMPS	Country

DICTIONARY Tables and Performance

When querying a DICTIONARY table, SAS launches a discovery process that gathers information that is pertinent to that table. Depending on the DICTIONARY table that is being queried, this discovery process can search libraries, open tables, and execute views. Unlike other SAS procedures and the DATA step, PROC SQL can mitigate this process by optimizing the query before the discovery process is launched. Therefore, although it is possible to access DICTIONARY table information with SAS procedures or the DATA step by using the SASHELP views, it is often more efficient to use PROC SQL instead.

Note: You cannot use data set options with DICTIONARY tables.

For example, the following programs produce the same result, but the PROC SQL step runs much faster because the WHERE clause is processed before the tables that are referenced by the SASHELP.VCOLUMN view are opened:

```
data mytable;
   set sashelp.vcolumn;
   where libname='WORK' and memname='SALES';
run;

proc sql;
   create table mytable as
      select * from sashelp.vcolumn
      where libname='WORK' and memname='SALES';
quit;
```

Note: SAS does not maintain DICTIONARY table information between queries. Each query of a DICTIONARY table launches a new discovery process.

If you are querying the same DICTIONARY table several times in a row, then you can get even faster performance by creating a temporary SAS data set (with the DATA step SET statement or the PROC SQL CREATE TABLE AS statement) with the information that you want and running your query against that data set.

When you query DICTIONARY.TABLES or SASHELP.VTABLE, all the tables and views in all the libraries that are assigned to the SAS session are opened to retrieve the requested information.

You can use a WHERE clause to help restrict which libraries are searched. However, the WHERE clause will not process most function calls such as UPCASE.

For example, if **`where UPCASE (libname) ='WORK'`** is used, the UPCASE function prevents the WHERE clause from optimizing this condition. All libraries that are assigned to the SAS session are searched. Searching all the libraries could cause an unexpected increase in search time, depending on the number of libraries that are assigned to the SAS session.

All librefs and SAS table names are stored in uppercase. If you supply values for LIBNAME and MEMNAME values in uppercase, and you remove the UPCASE function, the WHERE clause will be optimized and performance will be improved. In the previous example, the code would be changed to **`where libname='WORK'`**.

Note: If you query table information from a library that is assigned to an external database, and you use the LIBNAME statement PRESERVE_TAB_NAMES=YES option or the PRESERVE_COL_NAMES=YES option, and you provide the table or column name as it appears in the database, you do not need to use the UPCASE function.

Using SAS Data Set Options with PROC SQL

In PROC SQL, you can apply most of the SAS data set options, such as KEEP= and DROP=, to tables or SAS/ACCESS views any time you specify a table or SAS/ACCESS view. In the SQL procedure, SAS data set options that are separated by spaces are enclosed in parentheses. The data set options immediately follow the table or SAS/ACCESS view name. In the following PROC SQL step, the RENAME= data set option renames LNAME to LASTNAME for the STAFF1 table. The OBS= data set option restricts the number of rows that are read from STAFF1 to 15:

```
proc sql;
   create table
          staff1(rename=(lname=lastname)) as
      select *
         from staff(obs=15);
```

SAS data set options can be combined with SQL statement arguments. In the following PROC SQL step, the PW= data set option assigns a password to the TEST table, and the ALTER= data set option assigns an ALTER password to the STAFF1 table:

```
proc sql;
   create table test
      (a character, b numeric, pw=cat);
   create index staffidx on
      staff1 (lastname, alter=dog);
```

In this PROC SQL step, the PW= data set option assigns a password to the ONE table. The password is used when inserting a row and updating the table.

```
proc sql;
    create table one(pw=red, col1 num, col2 num, col3 num);
   quit;
```

```
proc sql;
   insert into one(pw=red, col1, col3)
   values(1, 3);
quit;
proc sql;
   update one(pw=red)
      set col2 = 22
         where col2 = . ;
quit;
```

You cannot use SAS data set options with DICTIONARY tables because DICTIONARY tables are read-only objects.

The only SAS data set options that you can use with PROC SQL views are data set options that assign and provide SAS passwords: READ=, WRITE=, ALTER=, and PW=.

For more information about SAS data set options, see *SAS Data Set Options: Reference*.

Using PROC SQL with the SAS Macro Facility

Overview of Using PROC SQL with the SAS Macro Facility

The macro facility is a programming tool that you can use to extend and customize SAS software. The macro facility reduces the amount of text that you must enter to perform common or repeated tasks and improves the efficiency and usefulness of your SQL programs.

The macro facility enables you to assign a name to character strings or groups of SAS programming statements. Thereafter, you can work with the names rather than with the text itself. For more information about the SAS macro facility, see *SAS Macro Language: Reference*.

Macro variables provide an efficient way to replace text strings in SAS code. The macro variables that you create and name are called user-defined macro variables. The macros variables that are defined by SAS are called automatic macro variables. PROC SQL produces six automatic macro variables (SQLOBS, SQLRC, SQLOOPS, SQLEXITCODE, SQLXRC, and SQLXMSG) to help you troubleshoot your programs. For more information, see “Using the PROC SQL Automatic Macro Variables” on page 157.

Creating Macro Variables in PROC SQL

Overview of Creating Macro Variables in PROC SQL

Other software vendors' SQL products allow the embedding of SQL into another language. References to variables (columns) of that language are termed host-variable references. They are differentiated from references to columns in tables by names that are prefixed with a colon. The host-variable stores the values of the object-items that are listed in the SELECT clause.

The only host language that is currently available in SAS is the macro language, which is part of Base SAS software. When a calculation is performed on a column's value, its result can be stored, using :macro-variable, in the macro facility. The result can then be referenced by that name in another PROC SQL query or SAS procedure. Host-variable can be used only in the outer query of a SELECT statement, not in a subquery. Host-variable cannot be used in a CREATE statement.

If the query produces more than one row of output, then the macro variable will contain only the value from the first row. If the query has no rows in its output, then the macro variable is not modified. If the macro variable does not exist yet, it will not be created. The PROC SQL macro variable SQLOBS contains the number of rows that are produced by the query.

Note: The SQLOBS automatic macro variable is assigned a value after the SQL SELECT statement executes.

Creating Macro Variables from the First Row of a Query Result

If you specify a single macro variable in the INTO clause, then PROC SQL assigns the variable the value from the first row only of the appropriate column in the SELECT list. In this example, &country1 is assigned the value from the first row of the Country column, and &barrels1 is assigned the value from the first row of the Barrels column. The NOPRINT option prevents PROC SQL from displaying the results of the query. The %PUT statement writes the contents of the macro variables to the SAS log.

```
libname sql 'SAS-library';

proc sql noprint;
   select country, barrels
      into :country1, :barrels1
      from sql.oilrsrvs;

%put &country1 &barrels1;
```

Log 5.5 Creating Macro Variables from the First Row of a Query Result

```
4   proc sql noprint;
5      select country, barrels
6         into :country1, :barrels1
7         from sql.oilrsrvs;
8
9   %put &country1 &barrels1;
Algeria                          9,200,000,000
NOTE: PROCEDURE SQL used:
      real time           0.12 seconds
```

Creating a Macro Variable from the Result of an Aggregate Function

A useful feature of macro variables is that they enable you to display data values in SAS titles. The following example prints a subset of the WORLDTEMPS table and lists the highest temperature in Canada in the title:

```
libname sql 'SAS-library';

proc sql outobs=12;
   reset noprint;
   select max(AvgHigh)
      into :maxtemp
      from sql.worldtemps
      where country = 'Canada';
reset print;
   title "The Highest Temperature in Canada: &maxtemp";
   select city, AvgHigh format 4.1
```

```
      from sql.worldtemps
      where country = 'Canada';
```

Note: You must use double quotation marks in the TITLE statement to resolve the reference to the macro variable.

Output 5.4 *Including a Macro Variable Reference in the Title*

The Highest Temperature in Canada: 80

City	AvgHigh
Montreal	77.0
Quebec	76.0
Toronto	80.0

Creating Multiple Macro Variables

You can create one new macro variable per row from the result of a SELECT statement. Use the keywords THROUGH, THRU, or a hyphen (-) in an INTO clause to create a range of macro variables.

Note: When you specify a range of macro variables, the SAS macro facility creates only the number of macro variables that are needed. For example, if you specify **`:var1-:var9999`** and only 55 variables are needed, only **`:var1-:var55`** is created. The SQLOBS automatic variable is useful if a subsequent part of your program needs to know how many variables were actually created. In this example, SQLOBS would have a value of 55.

This example assigns values to macro variables from the first four rows of the Name column and the first three rows of the Population column. The %PUT statements write the results to the SAS log.

```
libname sql 'SAS-library';

proc sql noprint;
   select name, Population
      into :country1 - :country4, :pop1 - :pop3
           from sql.countries;

%put &country1 &pop1;
%put &country2 &pop2;
%put &country3 &pop3;
%put &country4;
```

Log 5.6 *Creating Multiple Macro Variables*

```
4  proc sql noprint;
5     select name, Population
6        into :country1 - :country4, :pop1 - :pop3
7        from sql.countries;
8
9  %put &country1 &pop1;
Afghanistan 17070323
10  %put &country2 &pop2;
Albania 3407400
11  %put &country3 &pop3;
Algeria 28171132
12  %put &country4;
Andorra
```

Concatenating Values in Macro Variables

You can concatenate the values of one column into one macro variable. This form is useful for building a list of variables or constants. Use the SEPARATED BY keywords to specify a character to delimit the values in the macro variable.

This example assigns the first five values from the Name column of the COUNTRIES table to the &countries macro variable. The INOBS option limits PROC SQL to using the first five rows of the COUNTRIES table. A comma and a space are used to delimit the values in the macro variable.

```
libname sql 'SAS-library';

proc sql noprint inobs=5;
   select Name
      into :countries separated by ', '
      from sql.countries;

%put &countries;
```

Log 5.7 *Concatenating Values in Macro Variables*

```
4  proc sql noprint inobs=5;
5     select Name
6        into :countries separated by ', '
7           from sql.countries;
WARNING: Only 5 records were read from SQL.COUNTRIES due to INOBS= option.
8
9  %put &countries;
Afghanistan, Albania, Algeria, Andorra, Angola
```

The leading and trailing blanks are trimmed from the values before the macro variables are created. If you do not want the blanks to be trimmed, then add NOTRIM to the INTO clause. Here is the previous example with NOTRIM added:

```
libname sql 'SAS-library';

proc sql noprint inobs=5;
   select Name
      into :countries separated by ',' NOTRIM
      from sql.countries;
```

```
%put &countries;
```

Log 5.8 *Concatenating Values in Macro Variables*

```
1    proc sql noprint inobs=5;
2     select Name
3       into :countries separated by ',' NOTRIM
4       from sql.countries;
WARNING: Only 5 records were read from SQL.COUNTRIES due to INOBS= option.
5
6   %put &countries;
Afghanistan                         ,Albania                             ,Algeria
                           ,Andorra                             ,Angola
```

Defining Macros to Create Tables

Macros are useful as interfaces for table creation. You can use the SAS macro facility to help you create new tables and add rows to existing tables.

The following example creates a table that lists people to serve as referees for reviews of academic papers. No more than three people per subject are allowed in a table. The macro that is defined in this example checks the number of referees before it inserts a new referee's name into the table. The macro has two parameters: the referee's name and the subject matter of the academic paper.

```
libname sql 'SAS-library';

proc sql;
create table sql.referee
   (Name      char(15),
    Subject   char(15));

   /* define the macro */
%macro addref(name,subject);
%local count;

   /* are there three referees in the table? */
reset noprint;
   select count(*)
      into :count
      from sql.referee
      where subject="&subject";

%if &count ge 3 %then %do;
   reset print;
   title "ERROR: &name not inserted for subject - &subject..";
   title2 "        There are 3 referees already.";
   select * from sql.referee where subject="&subject";
   reset noprint;
   %end;

%else %do;
   insert into sql.referee(name,subject) values("&name","&subject");
   %put NOTE: &name has been added for subject - &subject..;
   %end;
```

```
%mend;
```

Submit the %ADDREF() macro with its two parameters to add referee names to the table. Each time you submit the macro, a message is written to the SAS log.

```
%addref(Conner,sailing);
%addref(Fay,sailing);
%addref(Einstein,relativity);
%addref(Smythe,sailing);
%addref(Naish,sailing);
```

Log 5.9 *Defining Macros to Create Tables*

```
34   %addref(Conner,sailing);
NOTE: 1 row was inserted into SQL.REFEREE.

NOTE: Conner has been added for subject - sailing.
35   %addref(Fay,sailing);
NOTE: 1 row was inserted into SQL.REFEREE.

NOTE: Fay has been added for subject - sailing.
36   %addref(Einstein,relativity);
NOTE: 1 row was inserted into SQL.REFEREE.

NOTE: Einstein has been added for subject - relativity.
37   %addref(Smythe,sailing);
NOTE: 1 row was inserted into SQL.REFEREE.

NOTE: Smythe has been added for subject - sailing.
38   %addref(Naish,sailing);
```

The output has a row added with each execution of the %ADDREF() macro. When the table contains three referee names, it is displayed in SAS output with the message that it can accept no more referees.

Output 5.5 *Result Table and Message Created with SAS Macro Language Interface*

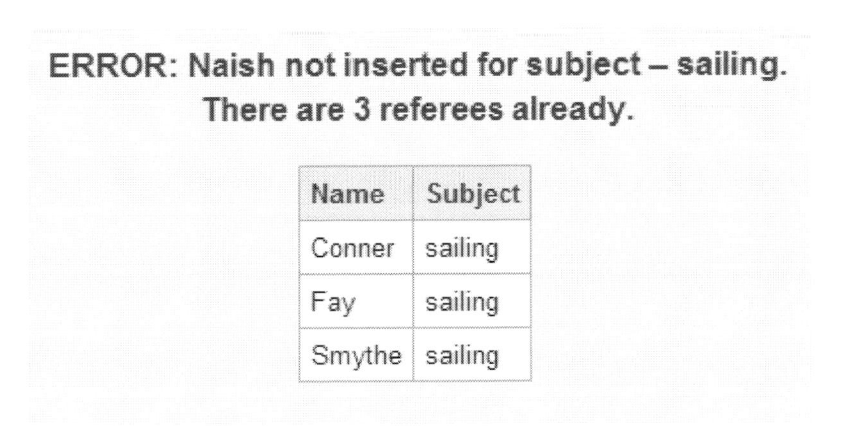

ERROR: Naish not inserted for subject – sailing.
There are 3 referees already.

Name	Subject
Conner	sailing
Fay	sailing
Smythe	sailing

Using the PROC SQL Automatic Macro Variables

PROC SQL sets up macro variables with certain values after it executes each statement. These macro variables can be tested inside a macro to determine whether to continue executing the PROC SQL step.

After each PROC SQL statement has executed, the following macro variables are updated with these values:

SQLEXITCODE
: contains the highest return code that occurred from some types of SQL insert failures. This return code is written to the SYSERR macro variable when PROC SQL terminates.

SQLOBS
: contains the number of rows that were processed by an SQL procedure statement. For example, the SQLOBS macro variable contains the number of rows that were formatted and displayed in SAS output by a SELECT statement or the number of rows that were deleted by a DELETE statement.

  When the NOPRINT option is specified, the value of the SQLOBS macro variable depends on whether an output table, single macro variable, macro variable list, or macro variable range is created:

  - If no output table, macro variable list, or macro variable range is created, then SQLOBS contains the value 1.
  - If an output table is created, then SQLOBS contains the number of rows in the output table.
  - If a single macro variable is created, then SQLOBS contains the value 1.
  - If a macro variable list or macro variable range is created, then SQLOBS contains the number of rows that are processed to create the macro variable list or range.

  If an SQL view is created, then SQLOBS contains the value 0.

  Note: The SQLOBS automatic macro variable is assigned a value after the SQL SELECT statement executes.

SQLOOPS
: contains the number of iterations that the inner loop of PROC SQL processes. The number of iterations increases proportionally with the complexity of the query. For more information, see "Limiting Iterations with the LOOPS= Option" on page 137 and LOOPS= in the *Base SAS Procedures Guide*.

SQLRC
: contains the following status values that indicate the success of the SQL procedure statement:

  0
  : PROC SQL statement completed successfully with no errors.

  4
  : PROC SQL statement encountered a situation for which it issued a warning. The statement continued to execute.

  8
  : PROC SQL statement encountered an error. The statement stopped execution at this point.

  12
  : PROC SQL statement encountered an internal error, indicating a bug in PROC SQL that should be reported to SAS Technical Support. These errors can occur only during compile time.

16
: PROC SQL statement encountered a user error. For example, this error code is used, when a subquery (that can return only a single value) evaluates to more than one row. These errors can be detected only during run time.

24
: PROC SQL statement encountered a system error. For example, this error is used, if the system cannot write to a PROC SQL table because the disk is full. These errors can occur only during run time.

28
: PROC SQL statement encountered an internal error, indicating a bug in PROC SQL that should be reported to SAS Technical Support. These errors can occur only during run time.

The value of SQLRC can vary based on the value of the PROC SQL statement UNDO_POLICY= option or the SQLUNDOPOLICY system option.

For example, the values for the SQLRC return code differ based on the value of the UNDO_POLICY= option or the SQLUNDOPOLICY system option if you attempt to insert duplicate values into an index that is defined using the CREATE UNIQUE INDEX statement:

- If you set the UNDO_POLICY= option or the SQLUNDOPOLICY system option to either REQUIRED or OPTIONAL, and you attempt to insert a duplicate index value, SAS creates and tries to maintain a copy of the table before and after updates are applied. SAS detects an error condition and supplies a return code to PROC SQL, which stops execution as soon as the error condition is received. SQLRC contains the value 24.
- If you set the UNDO_POLICY= option or the SQLUNDOPOLICY system option to NONE and you attempt to insert a duplicate index value, SAS does not create a before-and-after copy of the table. SAS does not detect an error condition and does not supply a return code to PROC SQL, which attempts to continue to process the updates. SQLRC contains the value 8.

SQLXMSG
: contains descriptive information and the DBMS-specific return code for the error that is returned by the pass-through facility.

Note: Because the value of the SQLXMSG macro variable can contain special characters (such as &, %, /, *, and ;), use the %SUPERQ macro function when printing the following value: **`%put %superq(sqlxmsg);`** For information about the %SUPERQ function, see *SAS Macro Language: Reference*.

SQLXRC
: contains the DBMS-specific return code that is returned by the pass-through facility.

Macro variables that are generated by PROC SQL follow the scoping rules for %LET. For more information about macro variable scoping, see *SAS Macro Language: Reference*.

Users of SAS/AF software can access these automatic macro variables in SAS Component Language (SCL) programs by using the SYMGET function. The following example uses the VALIDATE statement in a SAS/AF software application to check the syntax of a block of code. Before it issues the CREATE VIEW statement, the application checks that the view is accessible.

```
submit sql immediate;
   validate &viewdef;
end submit;
```

```
if symget('SQLRC') gt 4 then
   do;
      ... the view is not valid ...
   end;
else do;
   submit sql immediate;
      create view &viewname as &viewdef;
   end submit;
end;
```

The following example retrieves the data from the COUNTRIES table, but does not display the table because the NOPRINT option is specified in the PROC SQL statement. The %PUT macro language statement displays the three automatic macro variable values in the SAS log. For more information about the %PUT statement and the SAS macro facility, see *SAS Macro Language: Reference.*

```
libname sql 'SAS-library';

proc sql noprint;
   select * from sql.countries;
%put SQLOBS=*&sqlobs* SQLOOPS=*&sqloops* SQLRC=*&sqlrc*;
```

Log 5.10 *Using the PROC SQL Automatic Macro Variables*

```
SQLOBS=*1* SQLOOPS=*11* SQLRC=*0*
```

Notice that the value of SQLOBS is 1. When the NOPRINT option is used and no table or macro variables are created, SQLOBS returns a value of 1 because only one row is processed.

Note: You can use the _AUTOMATIC_ option in the %PUT statement to list the values of all automatic macro variables. The list depends on the SAS products that are installed at your site.

Formatting PROC SQL Output by Using the REPORT Procedure

SQL provides limited output formatting capabilities. Some SQL vendors add output formatting statements to their products to address these limitations. SAS has reporting tools that enhance the appearance of PROC SQL output.

For example, SQL cannot display only the first occurrence of a repeating value in a column in its output. The following example lists cities in the USCITYCOORDS table. Notice the repeating values in the State column.

```
libname sql 'SAS-library';

proc sql outobs=10;
   title 'US Cities';
   select State, City, latitude, Longitude
      from sql.uscitycoords
      order by state;
```

Output 5.6 *USCITYCOORDS Table Showing Repeating State Values*

US Cities

State	City	Latitude	Longitude
AK	Sitka	57	-135
AK	Anchorage	61	-150
AK	Nome	64	-165
AK	Juneau	58	-134
AL	Mobile	31	-88
AL	Montgomery	32	-86
AL	Birmingham	33	-87
AR	Hot Springs	34	-93
AR	Little Rock	35	-92
AZ	Flagstaff	35	-112

The following code uses PROC REPORT to format the output so that the state codes appear only once for each state group. A WHERE clause subsets the data so that the report lists the coordinates of cities in Pacific Rim states only. For more information about PROC REPORT, see the *Base SAS Procedures Guide*.

```
libname sql 'SAS-library';

proc sql noprint;
   create table sql.cityreport as
   select *
      from sql.uscitycoords
      group by state;

proc report data=sql.cityreport
            headline nowd
            headskip;
   title 'Coordinates of U.S. Cities in Pacific Rim States';
   column state city ('Coordinates' latitude longitude);
   define state / order format=$2. width=5 'State';
   define city / order format=$15. width=15 'City';
   define latitude / display format=4. width=8 'Latitude';
   define longitude / display format=4. width=9 'Longitude';
   where state='AK' or
         state='HI' or
         state='WA' or
         state='OR' or
         state='CA';
run;
```

Output 5.7 *PROC REPORT Output Showing the First Occurrence Only of Each State Value*

Coordinates of U.S. Cities in Pacific Rim States

		Coordinates	
State	**City**	**Latitude**	**Longitude**
AK	Anchorage	61	-150
	Juneau	58	-134
	Nome	64	-165
	Sitka	57	-135
CA	El Centro	32	-115
	Fresno	37	-120
	Long Beach	34	-118
	Los Angeles	34	-118
	Oakland	38	-122
	Sacramento	38	-121
	San Diego	33	-117
	San Francisco	38	-122
	San Jose	37	-122
HI	Honolulu	21	-158
OR	Baker	45	-118
	Eugene	44	-124
	Klamath Falls	42	-122
	Portland	45	-123
	Salem	45	-123
WA	Olympia	47	-123
	Seattle	47	-122
	Spokane	48	-117

Accessing a DBMS with SAS/ACCESS Software

Overview of Accessing a DBMS with SAS/ACCESS Software

SAS/ACCESS software for relational databases provides an interface between SAS software and data in other vendors' database management systems (DBMSs). SAS/ACCESS software provides dynamic access to DBMS data through the

SAS/ACCESS LIBNAME statement and the PROC SQL pass-through facility. The LIBNAME statement enables you to assign SAS librefs to DBMS objects such as schemas and databases. The pass-through facility enables you to interact with a DBMS by using its SQL syntax without leaving your SAS session.

It is recommended that you use the SAS/ACCESS LIBNAME statement to access your DBMS data because it is usually the fastest and most direct method of accessing DBMS data. The LIBNAME statement offers the following advantages:

- Significantly fewer lines of SAS code are required to perform operations in your DBMS. For example, a single LIBNAME statement establishes a connection to your DBMS, enables you to specify how your data is processed, and enables you to easily browse your DBMS tables in SAS.
- You do not need to know your DBMS's SQL language to access and manipulate your DBMS data. You can use SAS procedures, such as PROC SQL, or DATA step programming on any libref that references DBMS data. You can read, insert, update, delete, and append data, as well as create and drop DBMS tables by using normal SAS syntax.
- The LIBNAME statement provides more control over DBMS operations such as locking, spooling, and data type conversion through the many LIBNAME options and data set options.
- The LIBNAME engine optimizes the processing of joins and WHERE clauses by passing these operations directly to the DBMS to take advantage of the indexing and other processing capabilities of your DBMS.

An exception to this recommendation occurs when you need to use SQL that does not conform to the ANSI standard. The SAS/ACCESS LIBNAME statement accepts only ANSI-standard SQL, but the PROC SQL pass-through facility accepts all the extensions to SQL that are provided by your DBMS. Another advantage of this access method is that pass-through facility statements enable the DBMS to optimize queries when the queries have summary functions (such as AVG and COUNT), GROUP BY clauses, or columns that were created by expressions (such as the COMPUTED function).

For more information about SAS/ACCESS software, see *SAS/ACCESS for Relational Databases: Reference*.

Connecting to a DBMS by Using the LIBNAME Statement

Overview of Connecting to a DBMS by Using the LIBNAME Statement

Use the LIBNAME statement to read from and write to a DBMS object as if it were a SAS data set. After connecting to a DBMS table or view by using the LIBNAME statement, you can use PROC SQL to interact with the DBMS data.

For many DBMSs, you can directly access DBMS data by assigning a libref to the DBMS by using the SAS/ACCESS LIBNAME statement. Once you have associated a libref with the DBMS, you can specify a DBMS table in a two-level SAS name and work with the table like any SAS data set. You can also embed the LIBNAME statement in a PROC SQL view. For more information, see the CREATE VIEW Statement on page 234.

PROC SQL takes advantage of the capabilities of a DBMS by passing it certain operations whenever possible. For example, before implementing a join, PROC SQL checks to determine whether the DBMS can perform the join. If it can, then PROC SQL passes the join to the DBMS, which enhances performance by reducing data movement and translation. If the DBMS cannot perform the join, then PROC SQL processes the join. Using the SAS/ACCESS LIBNAME statement can often provide you with the performance benefits of the SQL procedure pass-through facility without writing DBMS-specific code.

Note: You can use the DBIDIRECTEXEC system option to send a PROC SQL CREATE TABLE AS SELECT statement or a DELETE statement directly to the database for execution, which could result in CPU and I/O performance improvement. For more information, see the SAS/ACCESS documentation for your DBMS.

To use the SAS/ACCESS LIBNAME statement, you must have SAS/ACCESS software installed for your DBMS. For more information about the SAS/ACCESS LIBNAME statement, see the SAS/ACCESS documentation for your DBMS.

Querying a DBMS Table

This example uses PROC SQL to query the Oracle table PAYROLL. The PROC SQL query retrieves all job codes and provides a total salary amount for each job code.

Note: By default, Oracle does not order the output results. To specify the order in which rows are displayed in the output results, you must use the ORDER BY clause in the SELECT statement.

```
libname mydblib oracle user=user-id password=password
        path=path-name schema=schema-name;

proc sql;
   select jobcode label='Jobcode',
          sum(salary) as total
          label='Total for Group'
          format=dollar11.2
      from mydblib.payroll
      group by jobcode;
quit;
```

Output 5.8 *Output from Querying a DBMS Table*

The SAS System

Jobcode	Total for Group
PT1	$543,264.00
BCK	$232,148.00
ME3	$296,875.00
NA2	$157,149.00
ME2	$498,076.00
TA3	$476,155.00
TA1	$249,492.00
PT3	$221,009.00
FA1	$253,433.00
SCP	$128,162.00
PT2	$879,252.00
FA2	$447,790.00
TA2	$671,499.00
ME1	$228,002.00
FA3	$230,537.00
NA1	$210,161.00

Creating a PROC SQL View of a DBMS Table

PROC SQL views are stored query expressions that read data values from their underlying files, which can include SAS/ACCESS views of DBMS data. While DATA step views of DBMS data can be used only to read the data, PROC SQL views of DBMS data can be used to update the underlying data if the following conditions are met:

- The PROC SQL view is based on only one DBMS table (or on a DBMS view that is based on only one DBMS table).
- The PROC SQL view has no calculated fields.

The following example uses the LIBNAME statement to connect to an ORACLE database, create a temporary PROC SQL view of the ORACLE table SCHEDULE, and print the view by using the PRINT procedure. The LIBNAME engine optimizes the processing of joins and WHERE clauses by passing these operations directly to the DBMS to take advantage of DBMS indexing and processing capabilities.

```
libname mydblib oracle user=user-id password=password
proc sql;
   create view LON as
   select flight, dates, idnum
      from mydblib.schedule
```

```
      where dest='LON';
quit;

proc print data=work.LON noobs;
run;
```

Output 5.9 *Output from the PRINT Procedure*

The SAS System

FLIGHT	DATES	IDNUM
219	01MAR1998:00:00:00	1407
219	01MAR1998:00:00:00	1777
219	01MAR1998:00:00:00	1103
219	01MAR1998:00:00:00	1125
219	01MAR1998:00:00:00	1350
219	01MAR1998:00:00:00	1332
219	02MAR1998:00:00:00	1407
219	02MAR1998:00:00:00	1118
219	02MAR1998:00:00:00	1132
219	02MAR1998:00:00:00	1135
219	02MAR1998:00:00:00	1441
219	02MAR1998:00:00:00	1332
219	03MAR1998:00:00:00	1428
219	03MAR1998:00:00:00	1442
219	03MAR1998:00:00:00	1130
219	03MAR1998:00:00:00	1411
219	03MAR1998:00:00:00	1115
219	03MAR1998:00:00:00	1332

Connecting to a DBMS by Using the SQL Procedure Pass-Through Facility

What Is the Pass-Through Facility?

The SQL procedure pass-through facility enables you to send DBMS-specific SQL statements directly to a DBMS for execution. The pass-through facility uses a SAS/ACCESS interface engine to connect to the DBMS. Therefore, you must have SAS/ACCESS software installed for your DBMS.

You submit SQL statements that are DBMS-specific. For example, you pass Transact-SQL statements to a Sybase database. The pass-through facility's basic syntax is the same for all the DBMSs. Only the statements that are used to connect to the DBMS and the SQL statements are DBMS-specific.

With the pass-through facility, you can perform the following tasks:

- Establish a connection with the DBMS by using a CONNECT statement and terminate the connection with the DISCONNECT statement.
- Send nonquery DBMS-specific SQL statements to the DBMS by using the EXECUTE statement.
- Retrieve data from the DBMS to be used in a PROC SQL query with the CONNECTION TO component in a SELECT statement's FROM clause.

You can use the pass-through facility statements in a query, or you can store them in a PROC SQL view. When a view is stored, any options that are specified in the corresponding CONNECT statement are also stored. Thus, when the PROC SQL view is used in a SAS program, SAS can automatically establish the appropriate connection to the DBMS.

For more information, see the CONNECT statement, the DISCONNECT statement, the EXECUTE statement, and the CONNECTION TO statement in Appendix 1, "SQL Macro Variables and System Options," on page 359, and the pass-through facility for relational databases in *SAS/ACCESS for Relational Databases: Reference*.

Note: SAS procedures that perform multipass processing cannot operate on PROC SQL views that store pass-through facility statements, because the pass-through facility does not allow reopening of a table after the first record has been retrieved. To work around this limitation, create a SAS data set from the view and use the SAS data set as the input data set.

Return Codes

As you use PROC SQL statements that are available in the pass-through facility, any errors are written to the SAS log. The return codes and messages that are generated by the pass-through facility are available to you through the SQLXRC and SQLXMSG macro variables. Both macro variables are described in "Using the PROC SQL Automatic Macro Variables" on page 157.

Pass-Through Example

In this example, SAS/ACCESS connects to an ORACLE database by using the alias **ora2**, selects all rows in the STAFF table, and displays the first 15 rows of data by using PROC SQL.

```
proc sql outobs=15;
   connect to oracle as ora2 (user=user-id password=password);
   select * from connection to ora2 (select lname, fname, state from staff);
   disconnect from ora2;
quit;
```

Output 5.10 *Output from the Pass-Through Facility Example*

The SAS System

LNAME	FNAME	STATE
ADAMS	GERALD	CT
ALIBRANDI	MARIA	CT
ALHERTANI	ABDULLAH	NY
ALVAREZ	MERCEDES	NY
ALVAREZ	CARLOS	NJ
BAREFOOT	JOSEPH	NJ
BAUCOM	WALTER	NY
BANADYGA	JUSTIN	CT
BLALOCK	RALPH	NY
BALLETTI	MARIE	NY
BOWDEN	EARL	CT
BRANCACCIO	JOSEPH	NY
BREUHAUS	JEREMY	NY
BRADY	CHRISTINE	CT
BREWCZAK	JAKOB	CT

Updating PROC SQL and SAS/ACCESS Views

You can update PROC SQL and SAS/ACCESS views by using the INSERT, DELETE, and UPDATE statements, under the following conditions:

- If the view accesses a DBMS table, then you must have been granted the appropriate authorization by the external database management system (for example, DB2). You must have installed the SAS/ACCESS software for your DBMS. For more information about SAS/ACCESS views, see the SAS/ACCESS interface guide for your DBMS.
- You can update only a single table through a view. The table cannot be joined to another table or linked to another table with a set-operator. The view cannot contain a subquery.
- You can update a column in a view by using the column's alias, but you cannot update a derived column—that is, a column that is produced by an expression. In the following example, you can update the column SS, but not WeeklySalary:

```
create view EmployeeSalaries as
   select Employee, SSNumber as SS,
          Salary/52 as WeeklySalary
          from employees;
```

- You cannot update a view that contains an ORDER BY.

Note: Beginning with SAS 9, PROC SQL views, the pass-through facility, and the SAS/ACCESS LIBNAME statement are the preferred ways to access relational DBMS data. SAS/ACCESS views are no longer recommended. You can convert existing SAS/ACCESS views to PROC SQL views by using the CV2VIEW procedure. For more information, see Chapter 32, "CV2VIEW Procedure" in *SAS/ACCESS for Relational Databases: Reference*.

Using the Output Delivery System with PROC SQL

The Output Delivery System (ODS) enables you to produce the output from PROC SQL in a variety of different formats such as PostScript, HTML, or list output. ODS defines the structure of the raw output from SAS procedures and from the SAS DATA step. The combination of data with a definition of its output structure is called an *output object*. Output objects can be sent to any of the various ODS destinations, which include listing, HTML, output, and printer. When new destinations are added to ODS, they automatically become available to PROC SQL, to all other SAS procedures that support ODS, and to the DATA step. For more information about ODS, see the *SAS Output Delivery System: User's Guide*.

The following example opens the HTML destination and specifies ODSOUT.HTM as the file that will contain the HTML output. The output from PROC SQL is sent to ODSOUT.HTM.

Note: This example uses filenames that might not be valid in all operating environments. To run the example successfully in your operating environment, you might need to change the file specifications.

Note: Some browsers require an extension of HTM or HTML on the filename.

```
libname sql 'SAS library';

ods html body='odsout.htm';
   proc sql outobs=12;
      title 'Coordinates of U.S. Cities';
      select *
         from sql.uscitycoords;
```

Output 5.11 *ODS HTML Output*

Coordinates of U.S. Cities

City	State	Latitude	Longitude
Albany	NY	43	-74
Albuquerque	NM	36	-106
Amarillo	TX	35	-102
Anchorage	AK	61	-150
Annapolis	MD	39	-77
Atlanta	GA	34	-84
Augusta	ME	44	-70
Austin	TX	30	-98
Baker	OR	45	-118
Baltimore	MD	39	-76

Measurements are in metric tons.

Chapter 6

Practical Problem-Solving with PROC SQL

Overview

This section shows you examples of solutions that PROC SQL can provide. Each example includes a statement of the problem to solve, background information that you must know to solve the problem, the PROC SQL solution code, and an explanation of how the solution works.

Computing a Weighted Average

Problem

You want to compute a weighted average of a column of values.

Background Information

There is one input table, called Sample, that contains the following data:

```
data Sample;
   do i=1 to 10;
      Value=2983*ranuni(135);
      Weight=33*rannor(579);
      if mod(i,2)=0 then Gender='M';
         else Gender='F';
      output;
   end;
   drop i;

proc print data=Sample;
  title 'Sample Data for Weighted Average';
run;
```

Output 6.1 *Sample Input Table for Weighted Averages*

Sample Data for Weighted Average

Obs	Value	Weight	Gender
1	2893.35	9.0868	F
2	56.13	26.2171	M
3	901.43	-4.0605	F
4	2942.68	-5.6557	M
5	621.16	24.3306	F
6	361.50	13.8971	M
7	2575.09	29.3734	F
8	2157.07	7.0687	M
9	690.73	-40.1271	F
10	2085.80	24.4795	M

Note that some of the weights are negative.

Solution

Use the following PROC SQL code to obtain weighted averages that are shown in the following output:

```
proc sql;
   title 'Weighted Averages from Sample Data';
   select Gender, sum(Value*Weight)/sum(Weight) as WeightedAverage
      from (select Gender, Value,
                   case
                      when Weight gt 0 then Weight
                      else 0
                   end as Weight
               from Sample)
      group by Gender;
```

Output 6.2 *PROC SQL Output for Weighted Averages*

Weighted Averages from Sample Data

Gender	WeightedAverage
F	1864.026
M	1015.91

How It Works

This solution uses an in-line view to create a temporary table that eliminates the negative data values in the Weight column. The in-line view is a query that performs the following tasks:

- selects the Gender and Value columns.
- uses a CASE expression to select the value from the Weight column. If Weight is greater than zero, then it is retrieved. If Weight is less than zero, then a value of zero is used in place of the Weight value.

```
(select Gender, Value,
        case
           when Weight>0 then Weight
           else 0
        end as Weight
    from Sample)
```

The first, or outer, SELECT statement in the query, performs the following tasks:

- selects the Gender column
- constructs a weighted average from the results that were retrieved by the in-line view

The weighted average is the sum of the products of Value and Weight divided by the sum of the Weights.

```
select Gender, sum(Value*Weight)/sum(Weight) as WeightedAverage
```

Finally, the query uses a GROUP BY clause to combine the data so that the calculation is performed for each gender.

```
group by Gender;
```

Comparing Tables

Problem

You have two copies of a table. One of the copies has been updated. You want to see which rows have been changed.

Background Information

There are two tables, the OLDSTAFF table and NEWSTAFF table. The NEWSTAFF table is a copy of OLDSTAFF. Changes have been made to NEWSTAFF. You want to find out what changes have been made.

Output 6.3 *Sample Input Tables for Table Comparison*

Old Staff Table

id	Last	First	Middle	Phone	Location
5463	Olsen	Mary	K.	661-0012	R2342
6574	Hogan	Terence	H.	661-3243	R4456
7896	Bridges	Georgina	W.	661-8897	S2988
4352	Anson	Sanford		661-4432	S3412
5674	Leach	Archie	G.	661-4328	S3533
7902	Wilson	Fran	R.	661-8332	R4454
0001	Singleton	Adam	O.	661-0980	R4457
9786	Thompson	Jack		661-6781	R2343

New Staff Table

id	Last	First	Middle	Phone	Location
5463	Olsen	Mary	K.	661-0012	R2342
6574	Hogan	Terence	H.	661-3243	R4456
7896	Bridges	Georgina	W.	661-2231	S2987
4352	Anson	Sanford		661-4432	S3412
5674	Leach	Archie	G.	661-4328	S3533
7902	Wilson	Fran	R.	661-8332	R4454
0001	Singleton	Adam	O.	661-0980	R4457
9786	Thompson	John	C.	661-6781	R2343
2123	Chen	Bill	W.	661-8099	R4432

Solution

To display only the rows that have changed in the new version of the table, use the EXCEPT set operator between two SELECT statements.

```
proc sql;
   title 'Updated Rows';
   select * from newstaff
   except
   select * from oldstaff;
```

Output 6.4 *Rows That Have Changed*

Updated Rows

id	Last	First	Middle	Phone	Location
2123	Chen	Bill	W.	661-8099	R4432
7896	Bridges	Georgina	W.	661-2231	S2987
9786	Thompson	John	C.	661-6781	R2343

How It Works

The EXCEPT operator returns rows from the first query that are not part of the second query. In this example, the EXCEPT operator displays only the rows that have been added or changed in the NEWSTAFF table.

Note: Any rows that were deleted from OLDSTAFF will not appear.

Overlaying Missing Data Values

Problem

You are forming teams for a new league by analyzing the averages of bowlers when they were members of other bowling leagues. When possible you will use each bowler's most recent league average. However, if a bowler was not in a league last year, then you will use the bowler's average from the prior year.

Background Information

There are two tables, LEAGUE1 and LEAGUE2, that contain bowling averages for last year and the prior year respectively. The structure of the tables is not identical because the data was compiled by two different secretaries. However, the tables do contain essentially the same type of data.

```
data league1;
input @1 Fullname $20. @21 Bowler $4. @29 AvgScore 3.;
cards;
Alexander Delarge   4224    164
John T Chance       4425
Jack T Colton       4264
                    1412    141
Andrew Shepherd     4189    185
;
```

```
data league2;
input @1 FirstName $10. @12 LastName $15. @28 AMFNo $4. @38 AvgScore 3.;
cards;
Alex       Delarge        4224      156
Mickey     Raymond        1412
                          4264      174
Jack       Chance         4425
Patrick    O'Malley       4118      164
;

proc sql;
title 'Bowling Averages from League1';
select * from league1;
title 'Bowling Averages from League2';
select * from league2;
```

Output 6.5 *Sample Input Tables for Overlaying Missing Values*

Bowling Averages from League1

Fullname	Bowler	AvgScore
Alexander Delarge	4224	164
John T Chance	4425	.
Jack T Colton	4264	.
	1412	141
Andrew Shepherd	4189	185

Bowling Averages from League2

FirstName	LastName	AMFNo	AvgScore
Alex	Delarge	4224	156
Mickey	Raymond	1412	.
		4264	174
Jack	Chance	4425	.
Patrick	O'Malley	4118	164

Solution

The following PROC SQL code combines the information from two tables, LEAGUE1 and LEAGUE2. The program uses all the values from the LEAGUE1 table, if available, and replaces any missing values with the corresponding values from the LEAGUE2 table. The results are shown in the following output.

```
proc sql;
   title "Averages from Last Year's League When Possible";
```

```
title2 "Supplemented when Available from Prior Year's League";
select coalesce(lastyr.fullname,trim(prioryr.firstname)
              ||' '||prioryr.lastname)as Name format=$26.,
       coalesce(lastyr.bowler,prioryr.amfno)as Bowler,
       coalesce(lastyr.avgscore,prioryr.avgscore)as Average format=8.
  from league1 as lastyr full join league2 as prioryr
       on lastyr.bowler=prioryr.amfno
  order by Bowler;
```

Output 6.6 *PROC SQL Output for Overlaying Missing Values*

Averages from Last Year's League When Possible
Supplemented when Available from Prior Year's League

Name	Bowler	Average
Mickey Raymond	1412	141
Patrick O'Malley	4118	164
Andrew Shepherd	4189	185
Alexander Delarge	4224	164
Jack T Colton	4264	174
John T Chance	4425	.

How It Works

This solution uses a full join to obtain all rows from LEAGUE1 as well as all rows from LEAGUE2. The program uses the COALESCE function on each column so that, whenever possible, there is a value for each column of a row. Using the COALESCE function on a list of expressions that is enclosed in parentheses returns the first nonmissing value that is found. For each row, the following code returns the AvgScore column from LEAGUE1 for Average:

```
coalesce(lastyr.avgscore,prioryr.avgscore) as Average format=8.
```

If this value of AvgScore is missing, then COALESCE returns the AvgScore column from LEAGUE2 for Average. If this value of AvgScore is missing, then COALESCE returns a missing value for Average.

In the case of the Name column, the COALESCE function returns the value of FullName from LEAGUE1 if it exists. If not, then the value is obtained from LEAGUE2 by using both the TRIM function and concatenation operators to combine the first name and last name columns:

```
trim(prioryr.firstname)||' '||prioryr.lastname
```

Finally, the table is ordered by Bowler. The Bowler column is the result of the COALESCE function.

```
coalesce(lastyr.bowler,prioryr.amfno)as Bowler
```

Because the value is obtained from either table, you cannot confidently order the output by either the value of Bowler in LEAGUE1 or the value of AMFNo in LEAGUE 2, but only by the value that results from the COALESCE function.

Computing Percentages within Subtotals

Problem

You want to analyze answers to a survey question to determine how each state responded. Then you want to compute the percentage of each answer that a given state contributed. For example, what percentage of all NO responses came from North Carolina?

Background Information

There is one input table, called SURVEY, that contains the following data (the first ten rows are shown):

```
data survey;
    input State $ Answer $ @@;
    datalines;
NY YES NY YES NY YES NY YES NY YES NY YES NY NO  NY NO  NY NO  NC YES
NC YES NC YES NC YES NC YES NC YES NC YES NC YES NC YES NC YES NC YES
NC YES NC YES NC YES NC YES NC YES NC YES NC YES NC YES NC YES NC NO
NC NO  NC NO  NC NO  NC NO  NC NO  NC NO  NC NO  NC NO  NC NO  NC NO
NC NO  NC NO  NC NO  NC NO  NC NO  NC NO  NC NO  NC NO  NC NO  NC NO
NC NO  NC NO  NC NO  PA YES PA YES PA YES PA YES PA YES PA YES PA YES
PA YES PA YES PA NO  PA NO  PA NO  PA NO  PA NO  PA NO  PA NO  PA NO
PA NO  PA NO  PA NO  PA NO  PA NO  PA NO  PA NO  PA NO  PA NO  PA NO
VA YES VA YES VA YES VA YES VA YES VA YES VA YES VA YES VA YES VA YES
VA YES VA YES VA YES VA YES VA YES VA YES VA YES VA YES VA YES VA NO
VA NO  VA NO  VA NO  VA NO  VA NO  VA NO  VA NO  VA NO  VA NO  VA NO
VA NO  VA NO  VA NO  VA NO  VA NO  VA NO
;

proc print data=survey(obs=10);
  title 'Sample Data for Subtotal Percentages';
run;
```

Output 6.7 *Input Table for Computing Subtotal Percentages (Partial Output)*

Sample Data for Subtotal Percentages

Obs	State	Answer
1	NY	YES
2	NY	YES
3	NY	YES
4	NY	YES
5	NY	YES
6	NY	YES
7	NY	NO
8	NY	NO
9	NY	NO
10	NC	YES

Solution

Use the following PROC SQL code to compute the subtotal percentages:

```
proc sql;
   title1 'Survey Responses';
   select survey.Answer, State, count(State) as Count,
          calculated Count/Subtotal as Percent format=percent8.2
   from survey,
        (select Answer, count(*) as Subtotal from survey
            group by Answer) as survey2
   where survey.Answer=survey2.Answer
   group by survey.Answer, State;
quit;
```

Output 6.8 *PROC SQL Output That Computes Percentages within Subtotals*

Survey Responses

Answer	State	Count	Percent
NO	NC	24	38.71%
NO	NY	3	4.84%
NO	PA	18	29.03%
NO	VA	17	27.42%
YES	NC	20	37.04%
YES	NY	6	11.11%
YES	PA	9	16.67%
YES	VA	19	35.19%

How It Works

This solution uses a subquery to calculate the subtotal counts for each answer. The code joins the result of the subquery with the original table and then uses the calculated state count as the numerator and the subtotal from the subquery as the denominator for the percentage calculation.

The query uses a GROUP BY clause to combine the data so that the calculation is performed for State within each answer.

```
group by survey.Answer, State;
```

Counting Duplicate Rows in a Table

Problem

You want to count the number of duplicate rows in a table and generate an output column that shows how many times each row occurs.

Background Information

There is one input table, called DUPLICATES, that contains the following data:

```
data Duplicates;
   input LastName $ FirstName $ City $ State $;
   datalines;
Smith John Richmond Virginia
Johnson Mary Miami Florida
Smith John Richmond Virginia
Reed Sam Portland Oregon
Davis Karen Chicago Illinois
Davis Karen Chicago Illinois
```

```
Thompson Jennifer Houston Texas
Smith John Richmond Virginia
Johnson Mary Miami Florida
;

proc print data=Duplicates;
  title 'Sample Data for Counting Duplicates';
run;
```

Output 6.9 *Sample Input Table for Counting Duplicates*

Sample Data for Counting Duplicates

Obs	LastName	FirstName	City	State
1	Smith	John	Richmond	Virginia
2	Johnson	Mary	Miami	Florida
3	Smith	John	Richmond	Virginia
4	Reed	Sam	Portland	Oregon
5	Davis	Karen	Chicago	Illinois
6	Davis	Karen	Chicago	Illinois
7	Thompson	Jennifer	Houston	Texas
8	Smith	John	Richmond	Virginia
9	Johnson	Mary	Miami	Florida

Solution

Use the following PROC SQL code to count the duplicate rows:

```
proc sql;
   title 'Duplicate Rows in DUPLICATES Table';
   select *, count(*) as Count
      from Duplicates
      group by LastName, FirstName, City, State
      having count(*) > 1;
```

Output 6.10 *PROC SQL Output for Counting Duplicates*

Duplicate Rows in DUPLICATES Table

LastName	FirstName	City	State	Count
Davis	Karen	Chicago	Illinois	2
Johnson	Mary	Miami	Florida	2
Smith	John	Richmond	Virginia	3

How It Works

This solution uses a query that performs the following:

- selects all columns
- counts all rows
- groups all of the rows in the Duplicates table by matching rows
- excludes the rows that have no duplicates

Note: You must include all of the columns in your table in the GROUP BY clause to find exact duplicates.

Expanding Hierarchical Data in a Table

Problem

You want to generate an output column that shows a hierarchical relationship among rows in a table.

Background Information

There is one input table, called EMPLOYEES, that contains the following data:

```
data Employees;
   input ID $ LastName $ FirstName $ Supervisor $;
   datalines;
1001 Smith John 1002
1002 Johnson Mary None
1003 Reed Sam None
1004 Davis Karen 1003
1005 Thompson Jennifer 1002
1006 Peterson George 1002
1007 Jones Sue 1003
1008 Murphy Janice 1003
1009 Garcia Joe 1002
;

proc print data=Employees;
  title 'Sample Data for Expanding a Hierarchy';
run;
```

Output 6.11 *Sample Input Table for Expanding a Hierarchy*

Sample Data for Expanding a Hierarchy

Obs	ID	LastName	FirstName	Supervisor
1	1001	Smith	John	1002
2	1002	Johnson	Mary	None
3	1003	Reed	Sam	None
4	1004	Davis	Karen	1003
5	1005	Thompson	Jennifer	1002
6	1006	Peterson	George	1002
7	1007	Jones	Sue	1003
8	1008	Murphy	Janice	1003
9	1009	Garcia	Joe	1002

You want to create output that shows the full name and ID number of each employee who has a supervisor, along with the full name and ID number of that employee's supervisor.

Solution

Use the following PROC SQL code to expand the data:

```
proc sql;
   title 'Expanded Employee and Supervisor Data';
   select A.ID label="Employee ID",
          trim(A.FirstName)||' '||A.LastName label="Employee Name",
          B.ID label="Supervisor ID",
          trim(B.FirstName)||' '||B.LastName label="Supervisor Name"
     from Employees A, Employees B
     where A.Supervisor=B.ID and A.Supervisor is not missing;
```

Output 6.12 *PROC SQL Output for Expanding a Hierarchy*

Expanded Employee and Supervisor Data

Employee ID	Employee Name	Supervisor ID	Supervisor Name
1001	John Smith	1002	Mary Johnson
1009	Joe Garcia	1002	Mary Johnson
1006	George Peterson	1002	Mary Johnson
1005	Jennifer Thompson	1002	Mary Johnson
1004	Karen Davis	1003	Sam Reed
1008	Janice Murphy	1003	Sam Reed
1007	Sue Jones	1003	Sam Reed

How It Works

This solution uses a self-join (reflexive join) to match employees and their supervisors. The SELECT clause assigns aliases of A and B to two instances of the same table and retrieves data from each instance. From instance A, the SELECT clause performs the following:

- selects the ID column and assigns it a label of **Employee ID**
- selects and concatenates the FirstName and LastName columns into one output column and assigns it a label of **Employee Name**

From instance B, the SELECT clause performs the following:

- selects the ID column and assigns it a label of **Supervisor ID**
- selects and concatenates the FirstName and LastName columns into one output column and assigns it a label of **Supervisor Name**

In both concatenations, the SELECT clause uses the TRIM function to remove trailing spaces from the data in the FirstName column, and then concatenates the data with a single space and the data in the LastName column to produce a single character value for each full name.

```
trim(A.FirstName)||' '||A.LastName label="Employee Name"
```

When PROC SQL applies the WHERE clause, the two table instances are joined. The WHERE clause conditions restrict the output to only those rows in table A that have a supervisor ID that matches an employee ID in table B. This operation provides a supervisor ID and full name for each employee in the original table, except for those who do not have a supervisor.

```
where A.Supervisor=B.ID and A.Supervisor is not missing;
```

Note: Although there are no missing values in the Employees table, you should check for and exclude missing values from your results to avoid unexpected results. For example, if there were an employee with a blank supervisor ID number and an employee with a blank ID, then they would produce an erroneous match in the results.

Summarizing Data in Multiple Columns

Problem

You want to produce a grand total of multiple columns in a table.

Background Information

There is one input table, called SALES, that contains the following data:

```
data Sales;
   input Salesperson $ January February March;
   datalines;
Smith 1000 650 800
Johnson 0 900 900
Reed 1200 700 850
Davis 1050 900 1000
Thompson 750 850 1000
Peterson 900 600 500
Jones 800 900 1200
Murphy 700 800 700
Garcia 400 1200 1150
;

proc print data=Sales;
  title 'Sample Data for Summarizing Data from Multiple Columns';
run;
```

Output 6.13 *Sample Input Table for Summarizing Data from Multiple Columns*

Sample Data for Summarizing Data from Multiple Columns

Obs	Salesperson	January	February	March
1	Smith	1000	650	800
2	Johnson	0	900	900
3	Reed	1200	700	850
4	Davis	1050	900	1000
5	Thompson	750	850	1000
6	Peterson	900	600	500
7	Jones	800	900	1200
8	Murphy	700	800	700
9	Garcia	400	1200	1150

You want to create output that shows the total sales for each month and the total sales for all three months.

Solution

Use the following PROC SQL code to produce the monthly totals and grand total:

```
proc sql;
   title 'Total First Quarter Sales';
   select sum(January)  as JanTotal,
          sum(February) as FebTotal,
          sum(March)    as MarTotal,
          sum(calculated JanTotal, calculated FebTotal,
              calculated MarTotal) as GrandTotal format=dollar10.
     from Sales;
```

Output 6.14 *PROC SQL Output for Summarizing Data from Multiple Columns*

Total First Quarter Sales

JanTotal	FebTotal	MarTotal	GrandTotal
6800	7500	8100	$22,400

How It Works

Recall that when you specify one column as the argument to an aggregate function, the values in that column are calculated. When you specify multiple columns, the values in each row of the columns are calculated. This solution uses the SUM function to calculate the sum of each month's sales, and then uses the SUM function a second time to total the monthly sums into one grand total.

```
sum(calculated JanTotal, calculated FebTotal,
    calculated MarTotal) as GrandTotal format=dollar10.
```

An alternative way to code the grand total calculation is to use nested functions:

```
sum(sum(January), sum(February), sum(March))
   as GrandTotal format=dollar10.
```

Creating a Summary Report

Problem

You have a table that contains detailed sales information. You want to produce a summary report from the detail table.

Background Information

There is one input table, called SALES, that contains detailed sales information. There is one record for each sale for the first quarter that shows the site, product, invoice number, invoice amount, and invoice date.

```
data sales;
  input Site $ Product $ Invoice $ InvoiceAmount InvoiceDate $;
  datalines;
V1009   VID010   V7679 598.5   980126
V1019   VID010   V7688 598.5   980126
V1032   VID005   V7771   1070  980309
V1043   VID014   V7780   1070  980309
V421    VID003   V7831   2000  980330
V421    VID010   V7832    750  980330
V570    VID003   V7762   2000  980302
V659    VID003   V7730   1000  980223
V783    VID003   V7815    750  980323
V985    VID003   V7733   2500  980223
V966    VID001   V5020   1167  980215
V98     VID003   V7750   2000  980223
;

proc sql;
  title 'Sample Data to Create Summary Sales Report';
  select * from sales;
quit;
```

Output 6.15 *Sample Input Table for Creating a Summary Report*

Sample Data to Create Summary Sales Report

Site	Product	Invoice	InvoiceAmount	InvoiceDate
V1009	VID010	V7679	598.5	980126
V1019	VID010	V7688	598.5	980126
V1032	VID005	V7771	1070	980309
V1043	VID014	V7780	1070	980309
V421	VID003	V7831	2000	980330
V421	VID010	V7832	750	980330
V570	VID003	V7762	2000	980302
V659	VID003	V7730	1000	980223
V783	VID003	V7815	750	980323
V985	VID003	V7733	2500	980223
V966	VID001	V5020	1167	980215
V98	VID003	V7750	2000	980223

You want to use this table to create a summary report that shows the sales for each product for each month of the quarter.

Solution

Use the following PROC SQL code to create a column for each month of the quarter, and use the summary function SUM in combination with the GROUP BY statement to accumulate the monthly sales for each product:

```
proc sql;
   title 'First Quarter Sales by Product';
   select Product,
          sum(Jan) label='Jan',
          sum(Feb) label='Feb',
          sum(Mar) label='Mar'
      from (select Product,
                   case
                      when substr(InvoiceDate,3,2)='01' then
                         InvoiceAmount end as Jan,
                   case
                      when substr(InvoiceDate,3,2)='02' then
                         InvoiceAmount end as Feb,
                   case
                      when substr(InvoiceDate,3,2)='03' then
                         InvoiceAmount end as Mar
               from work.sales)
      group by Product;
```

Output 6.16 *PROC SQL Output for a Summary Report*

First Quarter Sales by Product

Product	Jan	Feb	Mar
VID001	.	1167	.
VID003	.	5500	4750
VID005	.	.	1070
VID010	1197	.	750
VID014	.	.	1070

Note: Missing values in the matrix indicate that no sales occurred for that given product in that month.

How It Works

This solution uses an in-line view to create three temporary columns, Jan, Feb, and Mar, based on the month part of the invoice date column. The in-line view is a query that performs the following:

- selects the product column
- uses a CASE expression to assign the value of invoice amount to one of three columns, Jan, Feb, or Mar, depending on the value of the month part of the invoice date column

```
case
   when substr(InvoiceDate,3,2)='01' then
      InvoiceAmount end as Jan,
case
   when substr(InvoiceSate,3,2)='02' then
      InvoiceAmount end as Feb,
case
   when substr(InvoiceDate,3,2)='03' then
      InvoiceAmount end as Mar
```

The first, or outer, SELECT statement in the query performs the following:

- selects the product
- uses the summary function SUM to accumulate the Jan, Feb, and Mar amounts
- uses the GROUP BY statement to produce a line in the table for each product

Notice that dates are stored in the input table as strings. If the dates were stored as SAS dates, then the CASE expression could be written as follows:

```
case
   when month(InvoiceDate)=1 then
      InvoiceAmount end as Jan,
case
   when month(InvoiceDate)=2 then
      InvoiceAmount end as Feb,
case
```

```
when month(InvoiceDate)=3 then
   InvoiceAmount end as Mar
```

Creating a Customized Sort Order

Problem

You want to sort data in a logical, but not alphabetical, sequence.

Background Information

There is one input table, called CHORES, that contains the following data:

```
data chores;
   input Project $ Hours Season $;
   datalines;
weeding 48 summer
pruning 12 winter
mowing 36 summer
mulching 17 fall
raking 24 fall
raking 16 spring
planting 8 spring
planting 8 fall
sweeping 3 winter
edging 16 summer
seeding 6 spring
tilling 12 spring
aerating 6 spring
feeding 7 summer
rolling 4 winter
;


proc sql;
title 'Garden Chores';
select * from chores;
quit;
```

Output 6.17 *Sample Input Data for a Customized Sort*

Garden Chores

Project	Hours	Season
weeding	48	summer
pruning	12	winter
mowing	36	summer
mulching	17	fall
raking	24	fall
raking	16	spring
planting	8	spring
planting	8	fall
sweeping	3	winter
edging	16	summer
seeding	6	spring
tilling	12	spring
aerating	6	spring
feeding	7	summer
rolling	4	winter

You want to reorder this chore list so that all the chores are grouped by season, starting with spring and progressing through the year. Simply ordering by Season makes the list appear in alphabetical sequence: fall, spring, summer, winter.

Solution

Use the following PROC SQL code to create a new column, Sorter, that will have values of 1 through 4 for the seasons spring through winter. Use the new column to order the query, but do not select it to appear:

```
proc sql;
   title 'Garden Chores by Season in Logical Order';
   select Project, Hours, Season
      from (select Project, Hours, Season,
              case
                 when Season = 'spring' then 1
                 when Season = 'summer' then 2
                 when Season = 'fall' then 3
                 when Season = 'winter' then 4
                 else .
              end as Sorter
              from chores)
      order by Sorter;
```

Output 6.18 *PROC SQL Output for a Customized Sort Sequence*

Garden Chores by Season in Logical Order

Project	Hours	Season
tilling	12	spring
raking	16	spring
planting	8	spring
seeding	6	spring
aerating	6	spring
mowing	36	summer
feeding	7	summer
edging	16	summer
weeding	48	summer
raking	24	fall
mulching	17	fall
planting	8	fall
rolling	4	winter
pruning	12	winter
sweeping	3	winter

How It Works

This solution uses an in-line view to create a temporary column that can be used as an ORDER BY column. The in-line view is a query that performs the following:

- selects the Project, Hours, and Season columns
- uses a CASE expression to remap the seasons to the new column Sorter: spring to 1, summer to 2, fall to 3, and winter to 4

```
(select project, hours, season,
        case
           when season = 'spring' then 1
           when season = 'summer' then 2
           when season = 'fall' then 3
           when season = 'winter' then 4
           else .
        end as sorter
    from chores)
```

The first, or outer, SELECT statement in the query performs the following:

- selects the Project, Hours, and Season columns

- orders rows by the values that were assigned to the seasons in the Sorter column that was created with the in-line view

Notice that the Sorter column is not included in the SELECT statement. That causes a note to be written to the log indicating that you have used a column in an ORDER BY statement that does not appear in the SELECT statement. In this case, that is exactly what you wanted to do.

Conditionally Updating a Table

Problem

You want to update values in a column of a table, based on the values of several other columns in the table.

Background Information

There is one table, called INCENTIVES, that contains information about sales data. There is one record for each salesperson that includes a department code, a base pay rate, and sales of two products, gadgets and whatnots.

```
data incentives;
   input @1 Name $18. @20 Department $2. Payrate
            Gadgets Whatnots;
   datalines;
Lao Che            M2   8.00  10193  1105
Jack Colton        U2   6.00   9994  2710
Mickey Raymond     M1  12.00   6103  1930
Dean Proffit       M2  11.00   3000  1999
Antoinette Lily    E1  20.00   2203  4610
Sydney Wade        E2  15.00   4205  3010
Alan Traherne      U2   4.00   5020  3000
Elizabeth Bennett  E1  16.00  17003  3003
;
```

```
proc sql;
  title 'Sales Data for Incentives Program';
  select * from incentives;
quit;
```

Output 6.19 *Sample Input Data to Conditionally Change a Table*

Sales Data for Incentives Program

Name	Department	Payrate	Gadgets	Whatnots
Lao Che	M2	8	10193	1105
Jack Colton	U2	6	9994	2710
Mickey Raymond	M1	12	6103	1930
Dean Proffit	M2	11	3000	1999
Antoinette Lily	E1	20	2203	4610
Sydney Wade	E2	15	4205	3010
Alan Traherne	U2	4	5020	3000
Elizabeth Bennett	E1	16	17003	3003

You want to update the table by increasing each salesperson's payrate (based on the total sales of gadgets and whatnots) and taking into consideration some factors that are based on department code.

Specifically, anyone who sells over 10,000 gadgets merits an extra $5 per hour. Anyone selling between 5,000 and 10,000 gadgets also merits an incentive pay, but E Department salespersons are expected to be better sellers than those in the other departments, so their gadget sales incentive is $2 per hour compared to $3 per hour for those in other departments. Good sales of whatnots also entitle sellers to added incentive pay. The algorithm for whatnot sales is that the top level (level 1 in each department) salespersons merit an extra $.50 per hour for whatnot sales over 2,000, and level 2 salespersons merit an extra $1 per hour for sales over 2,000.

Solution

Use the following PROC SQL code to create a new value for the Payrate column. Actually Payrate is updated twice for each row, once based on sales of gadgets, and again based on sales of whatnots:

```
proc sql;
   update incentives
   set payrate = case
                    when gadgets > 10000 then
                       payrate + 5.00
                    when gadgets > 5000 then
                       case
                          when department in ('E1', 'E2') then
                             payrate + 2.00
                          else payrate + 3.00
                       end
                    else payrate
                 end;
   update incentives
   set payrate = case
                    when whatnots > 2000 then
```

```
                     case
                        when department in ('E2', 'M2', 'U2') then
                           payrate + 1.00
                        else payrate + 0.50
                     end
                  else payrate
               end;
title 'Adjusted Payrates Based on Sales of Gadgets and Whatnots';
select * from incentives;
```

Output 6.20 *PROC SQL Output for Conditionally Updating a Table*

Adjusted Payrates Based on Sales of Gadgets and Whatnots

Name	Department	Payrate	Gadgets	Whatnots
Lao Che	M2	13	10193	1105
Jack Colton	U2	10	9994	2710
Mickey Raymond	M1	15	6103	1930
Dean Proffit	M2	11	3000	1999
Antoinette Lily	E1	20.5	2203	4610
Sydney Wade	E2	16	4205	3010
Alan Traherne	U2	8	5020	3000
Elizabeth Bennett	E1	21.5	17003	3003

How It Works

This solution performs consecutive updates to the payrate column of the incentive table. The first update uses a nested case expression, first determining a bracket that is based on the amount of gadget sales: greater than 10,000 calls for an incentive of $5, between 5,000 and 10,000 requires an additional comparison. That is accomplished with a nested case expression that checks department code to choose between a $2 and $3 incentive.

```
update incentives
set payrate = case
                 when gadgets > 10000 then
                    payrate + 5.00
                 when gadgets > 5000 then
                    case
                       when department in ('E1', 'E2') then
                          payrate + 2.00
                       else payrate + 3.00
                    end
                 else payrate
              end;
```

The second update is similar, though simpler. All sales of whatnots over 2,000 merit an incentive, either $.50 or $1 depending on the department level, that again is accomplished by means of a nested case expression.

```
update incentives
   set payrate = case
                    when whatnots > 2000 then
                       case
                          when department in ('E2', 'M2', 'U2') then
                             payrate + 1.00
                          else payrate + 0.50
                       end
                    else payrate
                 end;
```

Updating a Table with Values from Another Table

Problem

You want to update the SQL.UNITEDSTATES table with updated population data.

Background Information

The SQL.NEWPOP table contains updated population data for some of the U.S. states.

```
libname sql 'SAS-library';

proc sql;
title 'Updated U.S. Population Data';
select state, population format=comma10. label='Population' from sql.newpop;
```

Output 6.21 *Table with Updated Population Data*

Updated U.S. Population Data

state	Population
Texas	20,851,820
Georgia	8,186,453
Washington	5,894,121
Arizona	5,130,632
Alabama	4,447,100
Oklahoma	3,450,654
Connecticut	3,405,565
Iowa	2,926,324
West Virginia	1,808,344
Idaho	1,293,953
Maine	1,274,923
New Hampshire	1,235,786
North Dakota	642,200
Alaska	626,932

Solution

Use the following PROC SQL code to update the population information for each state in the SQL.UNITEDSTATES table:

```
proc sql;
title 'UNITEDSTATES';
update sql.unitedstates as u
   set population=(select population from sql.newpop as n
        where u.name=n.state)
        where u.name in (select state from sql.newpop);
select Name format=$17., Capital format=$15.,
       Population, Area, Continent format=$13., Statehood format=date9.
   from sql.unitedstates;

/* use this code to generate output so you don't
   overwrite the sql.unitedstates table */
options ls=84;
proc sql outobs=10;
title 'UNITEDSTATES';
create table work.unitedstates as
   select * from sql.unitedstates;
update work.unitedstates as u
   set population=(select population from sql.newpop as n
```

```
            where u.name=n.state)
        where u.name in (select state from sql.newpop);
select Name format=$17., Capital format=$15.,
        Population, Area, Continent format=$13., Statehood format=date9.
 from work.unitedstates
;
```

Output 6.22 *SQL.UNITEDSTATES with Updated Population Data (Partial Output)*

UNITEDSTATES

Name	Capital	Population	Area	Continent	Statehood
Alabama	Montgomery	4447100	52423	North America	14DEC1819
Alaska	Juneau	626932	656400	North America	03JAN1959
Arizona	Phoenix	5130632	114000	North America	14FEB1912
Arkansas	Little Rock	2447996	53200	North America	15JUN1836
California	Sacramento	31518948	163700	North America	09SEP1850
Colorado	Denver	3601298	104100	North America	01AUG1876
Connecticut	Hartford	3405565	5500	North America	09JAN1788
Delaware	Dover	707232	2500	North America	07DEC1787
District of Colum	Washington	612907	100	North America	21FEB1871
Florida	Tallahassee	13814408	65800	North America	03MAR1845

How It Works

The UPDATE statement updates values in the SQL.UNITEDSTATES table (here with the alias U). For each row in the SQL.UNITEDSTATES table, the in-line view in the SET clause returns a single value. For rows that have a corresponding row in SQL.NEWPOP, this value is the value of the Population column from SQL.NEWPOP. For rows that do not have a corresponding row in SQL.NEWPOP, this value is missing. In both cases, the returned value is assigned to the Population column.

The WHERE clause ensures that only the rows in SQL.UNITEDSTATES that have a corresponding row in SQL.NEWPOP are updated, by checking each value of Name against the list of state names that is returned from the in-line view. Without the WHERE clause, rows that do not have a corresponding row in SQL.NEWPOP would have their Population values updated to missing.

Creating and Using Macro Variables

Problem

You want to create a separate data set for each unique value of a column.

Background Information

The SQL.FEATURES data set contains information about various geographical features around the world.

```
libname sql 'C:\Public\examples';

proc sql outobs=10;
title 'FEATURES';
 select Name format=$15., Type,Location format =$15.,Area,
        Height, Depth, Length
 from sql.features;
```

Output 6.23 *FEATURES (Partial Output)*

FEATURES

Name	Type	Location	Area	Height	Depth	Length
Aconcagua	Mountain	Argentina	.	22834	.	.
Amazon	River	South America	.	.	.	4000
Amur	River	Asia	.	.	.	2700
Andaman	Sea		218100	.	3667	.
Angel Falls	Waterfall	Venezuela	.	3212	.	.
Annapurna	Mountain	Nepal	.	26504	.	.
Aral Sea	Lake	Asia	25300	.	222	.
Ararat	Mountain	Turkey	.	16804	.	.
Arctic	Ocean		5105700	.	17880	.
Atlantic	Ocean		33420000	.	28374	.

Solution

To create a separate data set for each type of feature, you could go through the data set manually to determine all the unique values of Type, and then write a separate DATA step for each type (or a single DATA step with multiple OUTPUT statements). This approach is labor-intensive, error-prone, and impractical for large data sets. The following PROC SQL code counts the unique values of Type and puts each value in a separate macro variable. The SAS macro that follows the PROC SQL code uses these macro variables to create a SAS data set for each value. You do not need to know beforehand how many unique values there are or what the values are.

```
proc sql noprint;
   select count(distinct type)
      into :n
      from sql.features;
   select distinct type
```

```
      into :type1 - :type%left(&n)
      from sql.features;
quit;

%macro makeds;
   %do i=1 %to &n;
      data &&type&i (drop=type);
         set sql.features;
         if type="&&type&i";
      run;
   %end;
%mend makeds;
%makeds;
```

Log 6.1 *Log*

```
240  proc sql noprint;
241     select count(distinct type)
242        into :n
243        from sql.features;
244     select distinct type
245        into :type1 - :type%left(&n)
246        from sql.features;
247  quit;
NOTE: PROCEDURE SQL used (Total process time):
      real time           0.04 seconds
      cpu time            0.03 seconds


248
249  %macro makeds;
250     %do i=1 %to &n;
251        data &&type&i (drop=type);
252           set sql.features;
253           if type="&&type&i";
254        run;
255     %end;
256  %mend makeds;
257  %makeds;
NOTE: There were 74 observations read from the data set SQL.FEATURES.
NOTE: The data set WORK.DESERT has 7 observations and 6 variables.
NOTE: DATA statement used (Total process time):
      real time           1.14 seconds
      cpu time            0.41 seconds


NOTE: There were 74 observations read from the data set SQL.FEATURES.
NOTE: The data set WORK.ISLAND has 6 observations and 6 variables.
NOTE: DATA statement used (Total process time):
      real time           0.02 seconds
      cpu time            0.00 seconds


NOTE: There were 74 observations read from the data set SQL.FEATURES.
NOTE: The data set WORK.LAKE has 10 observations and 6 variables.
NOTE: DATA statement used (Total process time):
      real time           0.01 seconds
      cpu time            0.01 seconds


NOTE: There were 74 observations read from the data set SQL.FEATURES.
NOTE: The data set WORK.MOUNTAIN has 18 observations and 6 variables.
NOTE: DATA statement used (Total process time):
      real time           0.02 seconds
      cpu time            0.01 seconds


NOTE: There were 74 observations read from the data set SQL.FEATURES.
NOTE: The data set WORK.OCEAN has 4 observations and 6 variables.
NOTE: DATA statement used (Total process time):
      real time           0.01 seconds
      cpu time            0.01 seconds


NOTE: There were 74 observations read from the data set SQL.FEATURES.
NOTE: The data set WORK.RIVER has 12 observations and 6 variables.
NOTE: DATA statement used (Total process time):
      real time           0.02 seconds
      cpu time            0.02 seconds
```

```
NOTE: There were 74 observations read from the data set SQL.FEATURES.
NOTE: The data set WORK.SEA has 13 observations and 6 variables.
NOTE: DATA statement used (Total process time):
      real time           0.03 seconds
      cpu time            0.02 seconds

NOTE: There were 74 observations read from the data set SQL.FEATURES.
NOTE: The data set WORK.WATERFALL has 4 observations and 6 variables.
NOTE: DATA statement used (Total process time):
      real time           0.02 seconds
      cpu time            0.02 seconds
```

How It Works

This solution uses the INTO clause to store values in macro variables. The first SELECT statement counts the unique variables and stores the result in macro variable N. The second SELECT statement creates a range of macro variables, one for each unique value, and stores each unique value in one of the macro variables. Note the use of the %LEFT function, which trims leading blanks from the value of the N macro variable.

The MAKEDS macro uses all the macro variables that were created in the PROC SQL step. The macro uses a %DO loop to execute a DATA step for each unique value, writing rows that contain a given value of Type to a SAS data set of the same name. The Type variable is dropped from the output data sets.

For more information about SAS macros, see *SAS Macro Language: Reference*.

Using PROC SQL Tables in Other SAS Procedures

Problem

You want to show the average high temperatures in degrees Celsius for European countries on a map.

Background Information

The SQL.WORLDTEMPS table has average high and low temperatures for various cities around the world.

```
proc sql outobs=10;
title 'WORLDTEMPS';
 select City, Country,avghigh, avglow
  from sql.worldtemps
;
```

Output 6.24 *WORLDTEMPS (Partial Output)*

WORLDTEMPS

City	Country	AvgHigh	AvgLow
Algiers	Algeria	90	45
Amsterdam	Netherlands	70	33
Athens	Greece	89	41
Auckland	New Zealand	75	44
Bangkok	Thailand	95	69
Beijing	China	86	17
Belgrade	Yugoslavia	80	29
Berlin	Germany	75	25
Bogota	Colombia	69	43
Bombay	India	90	68

Solution

Use the following PROC SQL and PROC GMAP code to produce the map. You must license SAS/GRAPH software to use PROC GMAP.

```
options fmtsearch=(sashelp.mapfmts);

proc sql;
   create table extremetemps as
   select country, round((mean(avgHigh)-32)/1.8) as High,
      input(put(country,$glcsmn.), best.) as ID
   from sql.worldtemps
   where calculated id is not missing and country in
      (select name from sql.countries where continent='Europe')
   group by country;
quit;

proc gmap map=maps.europe data=extremetemps all;
   id id;
   block high / levels=3;
   title 'Average High Temperatures for European Countries';
   title2 'Degrees Celsius'
run;
quit;
```

Output 6.25 *PROC GMAP Output*

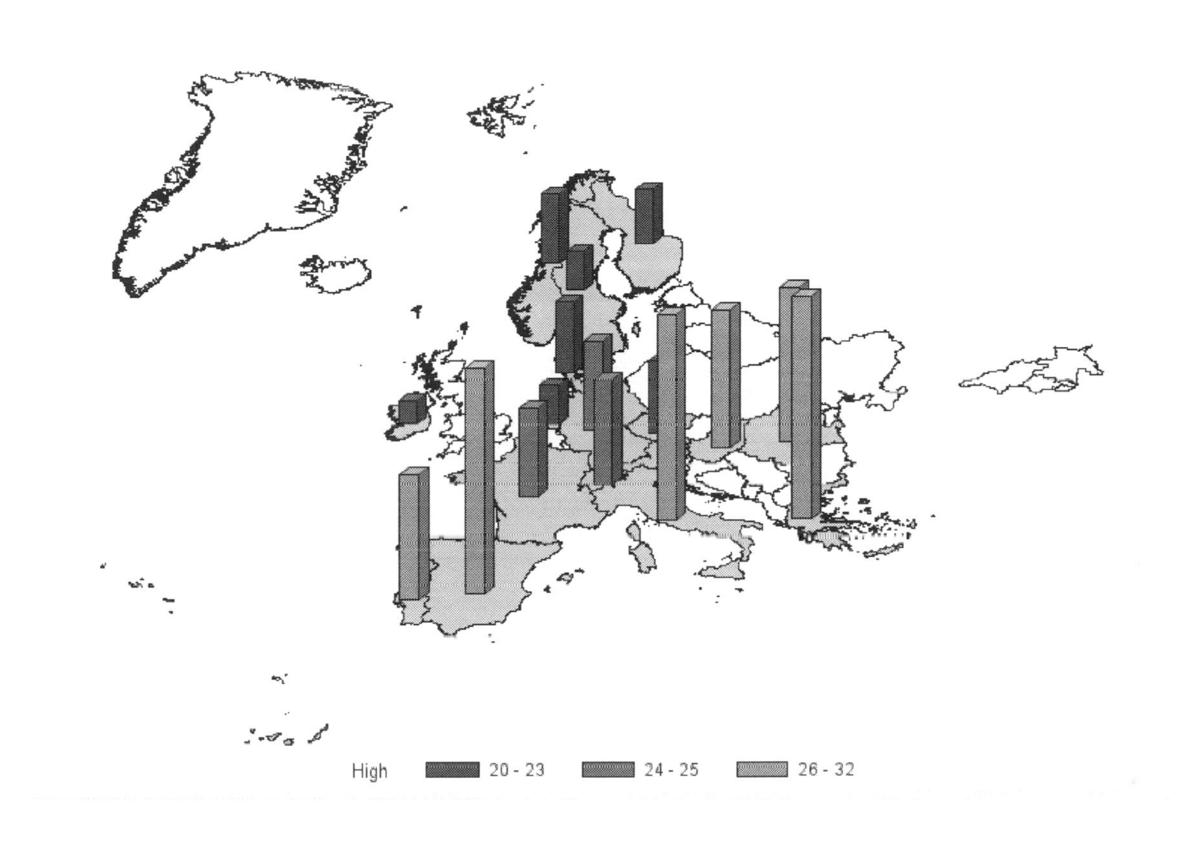

How It Works

The SAS system option FMTSEARCH= tells SAS to search in the SASHELP.MAPFMTS catalog for map-related formats. In the PROC SQL step, a temporary table is created with Country, High, and ID columns. The calculation **`round((mean(avgHigh)-32)/1.8)`** does the following:

1. For countries that are represented by more than one city, the mean of the cities' average high temperatures is used for that country.
2. That value is converted from degrees Fahrenheit to degrees Celsius.
3. The result is rounded to the nearest degree.

The PUT function uses the $GLCSMN. format to convert the country name to a country code. The INPUT function converts this country code, which is returned by the PUT function as a character value, into a numeric value that can be understood by the GMAP procedure. See *SAS Functions and CALL Routines: Reference* for details about the PUT and INPUT functions.

The WHERE clause limits the output to European countries by checking the value of the Country column against the list of European countries that is returned by the in-line view. Also, rows with missing values of ID are eliminated. Missing ID values could be produced if the $GLCSMN. format does not recognize the country name.

The GROUP BY clause is required so that the mean temperature can be calculated for each country rather than for the entire table.

The PROC GMAP step uses the ID variable to identify each country and places a block representing the High value on each country on the map. The ALL option ensures that countries (such as the United Kingdom in this example) that do not have High values are also drawn on the map. In the BLOCK statement, the LEVELS= option specifies how many response levels are used in the graph. For more information about the GMAP procedure, see *SAS/GRAPH: Reference*.

Part 2

SQL Procedure Reference

Chapter 7

SQL Procedure

Overview

What Is the SQL Procedure?

The SQL procedure implements Structured Query Language (SQL) for SAS. SQL is a standardized, widely used language that retrieves data from and updates data in tables and the views that are based on those tables.

The SAS SQL procedure enables you to

- retrieve and manipulate data that is stored in tables or views.
- create tables, views, and indexes on columns in tables.
- create SAS macro variables that contain values from rows in a query's result.
- add or modify the data values in a table's columns or insert and delete rows. You can also modify the table itself by adding, modifying, or dropping columns.
- send DBMS-specific SQL statements to a database management system (DBMS) and retrieve DBMS data.

The following figure summarizes the variety of source material that you can use with PROC SQL and what the procedure can produce.

Figure 7.1 *PROC SQL Input and Output*

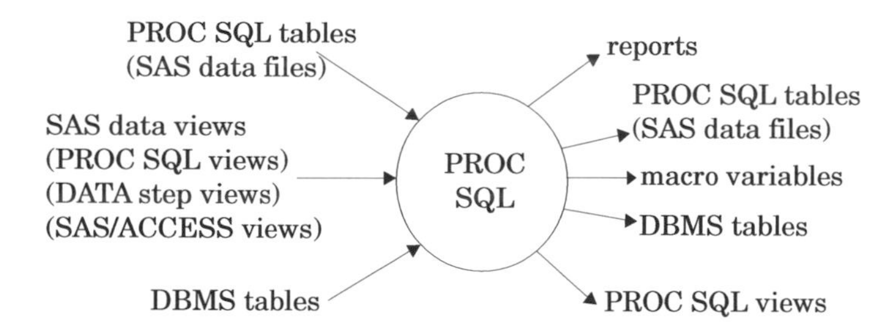

What Are PROC SQL Tables?

A PROC SQL table is synonymous with a SAS data file and has a member type of DATA. You can use PROC SQL tables as input into DATA steps and procedures.

You create PROC SQL tables from SAS data files, from SAS views, or from DBMS tables by using PROC SQL's pass-through facility or the SAS/ACCESS LIBNAME statement. The pass-through facility is described in "Connecting to a DBMS by Using the SQL Procedure Pass-Through Facility" on page 166. The SAS/ACCESS LIBNAME statement is described in "Connecting to a DBMS by Using the LIBNAME Statement" on page 163.

In PROC SQL terminology, a **row** in a table is the same as an **observation** in a SAS data file. A **column** is the same as a variable.

What Are Views?

A **SAS view** defines a virtual data set that is named and stored for later use. A view contains no data but describes or defines data that is stored elsewhere. There are three types of SAS views:

- PROC SQL views
- SAS/ACCESS views
- DATA step views.

You can refer to views in queries as if they were tables. The view derives its data from the tables or views that are listed in its FROM clause. The data that is accessed by a view is a subset or superset of the data that is in its underlying tables or views.

A PROC SQL view is a SAS data set of type VIEW that is created by PROC SQL. A PROC SQL view contains no data. It is a stored query expression that reads data values from its underlying files, which can include SAS data files, SAS/ACCESS views, DATA step views, other PROC SQL views, or DBMS data. When executed, a PROC SQL view's output can be a subset or superset of one or more underlying files.

SAS/ACCESS views and DATA step views are similar to PROC SQL views in that they are both stored programs of member type VIEW. SAS/ACCESS views describe data in DBMS tables from other software vendors. DATA step views are stored DATA step programs.

Note: Starting in SAS System 9, PROC SQL views, the pass-through facility, and the SAS/ACCESS LIBNAME statement are the preferred ways to access relational DBMS data; SAS/ACCESS views are no longer recommended. You can convert existing SAS/ACCESS views to PROC SQL views by using the CV2VIEW procedure. For more information, see Chapter 32, "CV2VIEW Procedure" in *SAS/ACCESS for Relational Databases: Reference*.

You can update data through a PROC SQL or SAS/ACCESS view with certain restrictions. See "Updating PROC SQL and SAS/ACCESS Views" on page 168.

You can use all types of views as input to DATA steps and procedures.

Note: In this chapter, the term **view** collectively refers to PROC SQL views, DATA step views, and SAS/ACCESS views, unless otherwise noted.

Note: When the contents of an SQL view are processed (by a DATA step or a procedure), the referenced data set must be opened to retrieve information about the variables that is not stored in the view. If that data set has a libref associated with it that is not defined in the current SAS code, then an error will result. You can avoid this error by specifying a USING clause in the CREATE VIEW statement. See CREATE VIEW Statement on page 234 for details.

Note: When you process PROC SQL views between a client and a server, getting the correct results depends on the compatibility between the client and server architecture. For more information, see "Accessing a SAS View" in Chapter 17 of *SAS/CONNECT User's Guide*.

SQL Procedure Coding Conventions

Because PROC SQL implements Structured Query Language, it works somewhat differently from other Base SAS procedures, as described here:

- When a PROC SQL statement is executed, PROC SQL continues to run until a QUIT statement, a DATA step, or another SAS procedure is executed. Therefore, you do not need to repeat the PROC SQL statement with each SQL statement. You need to repeat the PROC SQL statement only if you execute a QUIT statement, a DATA step, or another SAS procedure between SQL statements.
- SQL procedure statements are divided into clauses. For example, the most basic SELECT statement contains the SELECT and FROM clauses. Items within clauses are separated with commas in SQL, not with blanks as in other SAS code. For example, if you list three columns in the SELECT clause, then the columns are separated with commas.
- The SELECT statement, which is used to retrieve data, also automatically writes the output data to the Output window unless you specify the NOPRINT option in the PROC SQL statement. Therefore, you can display your output or send it to a list file without specifying the PRINT procedure.
- The ORDER BY clause sorts data by columns. In addition, tables do not need to be presorted by a variable for use with PROC SQL. Therefore, you do not need to use the SORT procedure with your PROC SQL programs.
- A PROC SQL statement runs when you submit it; you do not have to specify a RUN statement. If you follow a PROC SQL statement with a RUN statement, then SAS ignores the RUN statement and submits the statements as usual.

Syntax: SQL Procedure

Tips: Supports the Output Delivery System. For more information, see Chapter 3, “Output Delivery System: Basic Concepts,” in *SAS Output Delivery System: User's Guide*.

You can use any global statements. For more information, see Chapter 2, “Fundamental Concepts for Using Base SAS Procedures,” in *Base SAS Procedures Guide*.

You can use data set options any time a table name or view name is specified. For more information, see “Using SAS Data Set Options with PROC SQL” on page 151..

Regular type indicates the name of a component that is described in Chapter 9, “SQL Procedure Components,” on page 303 *view-name* indicates a SAS view of any type.

PROC SQL *<option(s)>*;

ALTER TABLE *table-name*

<ADD<CONSTRAINT>*constraint-clause*<, … *constraint-clause*>>

<ADD column-definition<, … column-definition>>

<DROP CONSTRAINT*constraint-name*<, … *constraint-name*>>

<DROP*column*<, … *column*>>

<DROP FOREIGN KEY*constraint-name*>

<DROP PRIMARY KEY>

<MODIFY column-definition<, … column-definition>>

;

CREATE <UNIQUE> **INDEX** *index-name*

ON *table-name* (*column* <, … *column*>);

CREATE TABLE *table-name*

(*column-specification*<, …*column-specification* | *constraint-specification*>)

;

CREATE TABLE *table-name* **LIKE** *table-name2*;

CREATE TABLE *table-name* **AS** query-expression

<ORDER BY*order-by-item*<, … *order-by-item*>>;

CREATE VIEW *proc-sql-view* **AS** query-expression

<ORDER BY*order-by-item*<, … *order-by-item*>>

<USING*libname-clause*<, … *libname-clause*>> ;

DELETE

FROM *table-name|proc-sql-view |sas/access-view* <AS *alias*>

<WHERE sql-expression>;

DESCRIBE TABLE *table-name* <, … *table-name*>;

DESCRIBE VIEW *proc-sql-view* <, … *proc-sql-view*>;

DESCRIBE TABLE CONSTRAINTS *table-name* <, … *table-name*>;

DROP INDEX *index-name* <, … *index-name*>
FROM *table-name*;

DROP TABLE *table-name* <, … *table-name*>;

DROP VIEW *view-name* <, … *view-name*>;

INSERT INTO *table-name|sas/access-view|proc-sql-view* <(*column*<, … *column*>)>

SET *column*=sql-expression
<, … *column*=sql-expression>

<SET*column*=sql-expression<, … *column*=sql-expression>>;

INSERT INTO *table-name|sas/access-view|proc-sql-view* <(*column*<, … *column*>)>

VALUES (*value* <, … *value*>)
<… VALUES (*value*<, … *value*>)>;

INSERT INTO *table-name|sas/access-view|proc-sql-view*
<(*column*<, …*column*>)> query-expression;

RESET *<option(s)>*;

SELECT <DISTINCT> *object-item* <, …*object-item*>
<INTO*macro-variable-specification*<, … *macro-variable-specification*>>
FROM *from-list*
<WHERE sql-expression>
<GROUP BY*group-by-item*<, … *group-by-item*>>
<HAVING sql-expression>
<ORDER BY*order-by-item*<, … *order-by-item*>>;
UPDATE *table-name|sas/access-view|proc-sql-view* <AS*alias*>
SET *column*=sql-expression
<, … *column*=sql-expression>
<SET*column*=sql-expression<, … *column*=sql-expression>>
<WHERE sql-expression>;
VALIDATE query-expression;

To connect to a DBMS and send it a DBMS-specific nonquery SQL statement, use this form:

PROC SQL;
CONNECT TO *dbms-name* <AS *alias*>
<(*connect-statement-argument-1=value*<… *connect-statement-argument-n=value*>)>
<(*database-connection-argument-1=value*<… *database-connection-argument-n=value*>)>;
EXECUTE (*dbms-SQL-statement*)
BY *dbms-name|alias*;
<DISCONNECT FROM*dbms-name|alias*;>
<QUIT;>

To connect to a DBMS and query the DBMS data, use this form:

PROC SQL;
CONNECT TO *dbms-name* <AS *alias*>
<(*connect-statement-argument-1=value*<… *connect-statement-argument-n=value*>)>
<(*database-connection-argument-1=value*<… *database-connection-argument-n=value*>)>;
SELECT *column-list*
FROM CONNECTION TO *dbms-name|alias*
(*dbms-query*)
optional PROC SQL clauses;
<DISCONNECT FROM*dbms-name|alias*;>
<QUIT;>

Statement	Task
PROC SQL Statement	Create, maintain, retrieve, and update data in tables and views that are based on these tables
ALTER TABLE Statement	Modify, add, or drop columns
CONNECT Statement	Establish a connection with a DBMS
CREATE INDEX Statement	Create an index on a column
CREATE TABLE Statement	Create a PROC SQL table

Statement	Task
CREATE VIEW Statement	Create a PROC SQL view
DELETE Statement	Delete rows
DESCRIBE Statement	Display a definition of a table or view
DISCONNECT Statement	Terminate the connection with a DBMS
DROP Statement	Delete tables, views, or indexes
EXECUTE Statement	Send a DBMS-specific nonquery SQL statement to a DBMS
INSERT Statement	Add rows
RESET Statement	Reset options that affect the procedure environment without restarting the procedure
SELECT Statement	Select and execute rows
UPDATE Statement	Modify values
VALIDATE Statement	Verify the accuracy of your query

PROC SQL Statement

PROC SQL Statement

Syntax

PROC SQL *<option(s)>*;

Summary of Optional Arguments

Control execution

CONSTDATETIME|NOCONSTDATETIME

DQUOTE=ANSI|SAS

ERRORSTOP|NOERRORSTOP

EXEC|NOEXEC

EXITCODE

INOBS=*n*

IPASSTHRU|NOIPASSTHRU

LOOPS=*n*

NOCONSTDATETIME

NOERRORSTOP

NOEXEC

NOIPASSTHRU
NOPROMPT
NOREMERGE
NOSTIMER
NOTHREADS
OUTOBS=*n*
PROMPT|NOPROMPT
REDUCEPUT=ALL|NONE|DBMS|BASE
REDUCEPUTOBS=n
REDUCEPUTVALUES=n
REMERGE|NOREMERGE
STIMER|NOSTIMER
STOPONTRUNC
THREADS|NOTHREADS
UNDO_POLICY=NONE|OPTIONAL|REQUIRED

Control output

BUFFERSIZE=n|nK|nM|nG
DOUBLE|NODOUBLE
FEEDBACK|NOFEEDBACK
FLOW<=*n* <*m*>>|NOFLOW
NODOUBLE
NOFEEDBACK
NOFLOW
NONUMBER
NOPRINT
NOSORTMSG
NOWARNRECURS
NUMBER|NONUMBER
PRINT|NOPRINT
SORTMSG|NOSORTMSG
SORTSEQ=*sort-table*
WARNRECURS|NOWARNRECURS

Optional Arguments

BUFFERSIZE=n|nK|nM|nG

specifies the permanent buffer page size for the output in multiples of 1 (bytes), 1024 (kilobytes), 1,048,576 (megabytes), or 1,073,741,824 (gigabytes). For example, a value of 65536 specifies a page size of `65536` bytes and a value of `64k` specifies a page size of 65536 bytes.

BUFFERSIZE can also be specified in a RESET statement for use in particular queries.

Default: 0, which causes SAS to use the minimum optimal page size for the operating environment.

CONSTDATETIME|NOCONSTDATETIME

specifies whether the SQL procedure replaces references to the DATE, TIME, DATETIME, and TODAY functions in a query with their equivalent constant values before the query executes. Computing these values once ensures consistency of results when the functions are used multiple times in a query or when the query executes the functions close to a date or time boundary.

When the NOCONSTDATETIME option is set, PROC SQL evaluates these functions in a query each time it processes an observation.

Default: CONSTDATETIME

Interaction: If both the CONSTDATETIME option and the REDUCEPUT= option on page 220 are specified, PROC SQL replaces the DATE, TIME, DATETIME, and TODAY functions with their respective values in order to determine the PUT function value before the query executes.

Tip: Alternatively, you can set the SQLCONSTDATETIME system option. The value that is specified in the SQLCONSTDATETIME system option is in effect for all SQL procedure statements, unless the PROC SQL CONSTDATETIME option is set. The value of the CONSTDATETIME option takes precedence over the SQLCONSTDATETIME system option. The RESET statement can also be used to set or reset the CONSTDATETIME option. However, changing the value of the CONSTDATETIME option does not change the value of the SQLCONSTDATETIME system option. For more information, see the "SQLCONSTDATETIME System Option" on page 359.

DOUBLE|NODOUBLE

double-spaces the report.

Default: NODOUBLE

Example: "Example 5: Combining Two Tables" on page 254

DQUOTE=ANSI|SAS

specifies whether PROC SQL treats values within double quotation marks (" ") as variables or strings. With DQUOTE=ANSI, PROC SQL treats a quoted value as a variable. This feature enables you to use the following as table names, column names, or aliases:

- reserved words such as AS, JOIN, GROUP, and so on
- DBMS names and other names that are not normally permissible in SAS.

The quoted value can contain any character.

With DQUOTE=SAS, values within double quotation marks are treated as strings.

Default: SAS

ERRORSTOP|NOERRORSTOP

specifies whether PROC SQL stops executing if it encounters an error. In a batch or noninteractive session, ERRORSTOP instructs PROC SQL to stop executing the statements but to continue checking the syntax after it has encountered an error.

NOERRORSTOP instructs PROC SQL to execute the statements and to continue checking the syntax after an error occurs.

Default: NOERRORSTOP in an interactive SAS session; ERRORSTOP in a batch or noninteractive session

Interaction: This option is useful only when the EXEC option is in effect.

Tips:

ERRORSTOP has an effect only when SAS is running in the batch or noninteractive execution mode.

NOERRORSTOP is useful if you want a batch job to continue executing SQL procedure statements after an error is encountered.

EXEC|NOEXEC
specifies whether a statement should be executed after its syntax is checked for accuracy.

Default: EXEC

Tip: NOEXEC is useful if you want to check the syntax of your SQL statements without executing the statements.

See: ERRORSTOP on page 217

EXITCODE
specifies whether PROC SQL clears an error code for any SQL statement. Error codes are assigned to the SQLEXITCODE macro variable.

Default: 0

Tip: The exit code can be reset to the default value between PROC SQL statements with the RESET Statement on page 242.

See: "Using the PROC SQL Automatic Macro Variables" on page 157

FEEDBACK|NOFEEDBACK
specifies whether PROC SQL displays, in the SAS log, PROC SQL statements after view references are expanded or certain other transformations of the statement are made.

This option has the following effects:

- Any asterisk (for example, `SELECT *`) is expanded into the list of qualified columns that it represents.
- Any PROC SQL view is expanded into the underlying query.
- Macro variables are resolved.
- Parentheses are shown around all expressions to further indicate their order of evaluation.
- Comments are removed.

Default: NOFEEDBACK

FLOW<=*n* <*m*>>|NOFLOW
specifies that character columns longer than *n* are flowed to multiple lines. PROC SQL sets the column width at *n* and specifies that character columns longer than *n* are flowed to multiple lines. When you specify FLOW=*n m*, PROC SQL floats the width of the columns between these limits to achieve a balanced layout. Specifying FLOW without arguments is equivalent to specifying FLOW=12 200.

Default: NOFLOW

INOBS=*n*
restricts the number of rows (observations) that PROC SQL retrieves from any single source.

Tip: This option is useful for debugging queries on large tables.

IPASSTHRU|NOIPASSTHRU
specifies whether implicit pass through is enabled or disabled.

Implicit pass through is enabled when PROC SQL is invoked. You can disable it for a query or series of queries. The primary reasons that you might want to disable implicit pass through are as follows:

- DBMSs use SQL2 semantics for NULL values, which behave somewhat differently than SAS missing values.
- PROC SQL might do a better job of query optimization.

Default: IPASSTHRU

See: The documentation on the pass-through facility for your DBMS in *SAS/ACCESS for Relational Databases: Reference*.

LOOPS=*n*

restricts PROC SQL to *n* iterations through its inner loop. You use the number of iterations reported in the SQLOOPS macro variable (after each SQL statement is executed) to discover the number of loops. Set a limit to prevent queries from consuming excessive computer resources. For example, joining three large tables without meeting the join-matching conditions could create a huge internal table that would be inefficient to execute.

See: "Using the PROC SQL Automatic Macro Variables" on page 157

NOCONSTDATETIME

See CONSTDATETIME|NOCONSTDATETIME on page 217

NODOUBLE

See DOUBLE|NODOUBLE on page 217

NOERRORSTOP

See ERRORSTOP|NOERRORSTOP on page 217

NOEXEC

See EXEC|NOEXEC on page 218

NOFEEDBACK

See FEEDBACK|NOFEEDBACK on page 218

NOFLOW

See FLOW<–n <m>>|NOFLOW on page 218.

NOIPASSTHRU

See IPASSTHRU|NOIPASSTHRU on page 218

NONUMBER

See NUMBER|NONUMBER on page 220

NOPRINT

See PRINT|NOPRINT on page 220

NOPROMPT

See PROMPT|NOPROMPT on page 220

NOREMERGE

See REMERGE|NOREMERGE on page 222

NOSORTMSG

See SORTMSG|NOSORTMSG on page 222

NOSTIMER

See STIMER|NOSTIMER on page 222

NOTHREADS

See THREADS|NOTHREADS on page 222

NOWARNRECURS

See WARNRECURS|NOWARNRECURS on page 223

NUMBER|NONUMBER
specifies whether the SELECT statement includes a column called ROW, which is the row (or observation) number of the data as the rows are retrieved.

Default: NONUMBER

Example: "Example 4: Joining Two Tables" on page 251

OUTOBS=*n*
restricts the number of rows (observations) in the output. For example, if you specify OUTOBS=10 and insert values into a table using a query expression, then the SQL procedure inserts a maximum of 10 rows. Likewise, OUTOBS=10 limits the output to 10 rows.

PRINT|NOPRINT
specifies whether the output from a SELECT statement is printed.

Default: PRINT

Interaction: NOPRINT affects the value of the SQLOBS automatic macro variable. For more information, see "Using the PROC SQL Automatic Macro Variables" on page 157.

Tip: NOPRINT is useful when you are selecting values from a table into macro variables and do not want anything to be displayed.

PROMPT|NOPROMPT
modifies the effect of the INOBS=, OUTOBS=, and LOOPS= options. If you specify the PROMPT option and reach the limit specified by INOBS=, OUTOBS=, or LOOPS=, then PROC SQL prompts you to stop or continue. The prompting repeats if the same limit is reached again.

Default: NOPROMPT

REDUCEPUT=ALL|NONE|DBMS|BASE
specifies the engine type to use to optimize a PUT function in a query. The PUT function is replaced with a logically equivalent expression. The engine type can be one of the following values:

ALL
specifies to consider the optimization of all PUT functions, regardless of the engine that is used by the query to access the data.

NONE
specifies to not optimize any PUT function.

DBMS
specifies to consider the optimization of all PUT functions in a query performed by a SAS/ACCESS engine.

Requirement: The first argument to the PUT function must be a variable that is obtained by a table. The table must be accessed using a SAS/ACCESS engine.

BASE
specifies to consider the optimization of all PUT functions in a query performed by a SAS/ACCESS engine or a Base SAS engine.

Default: DBMS

Interactions:
If both the REDUCEPUT= option and the CONSTDATETIME option are specified, PROC SQL replaces the DATE, TIME, DATETIME, and TODAY functions with their respective values to determine the PUT function value before the query executes.

If the query also contains a WHERE or HAVING clause, the evaluation of the WHERE or HAVING clause is simplified.

Tip: Alternatively, you can set the SQLREDUCEPUT= system option. The value that is specified in the SQLREDUCEPUT= system option is in effect for all SQL procedure statements, unless the REDUCEPUT= option is set. The value of the REDUCEPUT= option takes precedence over the SQLREDUCEPUT= system option. The RESET statement can also be used to set or reset the REDUCEPUT= option. However, changing the value of the REDUCEPUT= option does not change the value of the SQLREDUCEPUT= system option. For more information, see the "SQLREDUCEPUT= System Option" on page 364.

REDUCEPUTOBS=n

when the REDUCEPUT= option is set to DBMS, BASE, or ALL, specifies the minimum number of observations that must be in a table for PROC SQL to consider optimizing the PUT function in a query.

Default: 0, which indicates that there is no minimum number of observations in a table for PROC SQL to optimize the PUT function.

Range: $0 - 2^{63}-1$, or approximately 9.2 quintillion

Requirement: *n* must be an integer

Interaction: The REDUCEPUTOBS= option works only for DBMSs that record the number of observations in a table. If your DBMS does not record the number of observations, but you create row counts on your table, the REDUCEPUTOBS= option will work.

Tip: Alternatively, you can set the SQLREDUCEPUTOBS= system option. The value that is specified in the SQLREDUCEPUTOBS= system option is in effect for all SQL procedure statements, unless the REDUCEPUTOBS= option is set. The value of the REDUCEPUTOBS= option takes precedence over the SQLREDUCEPUTOBS= system option. The RESET statement can also be used to set or reset the REDUCEPUTOBS= option. However, changing the value of the REDUCEPUTOBS= option does not change the value of the SQLREDUCEPUTOBS= system option. For more information, see the "SQLREDUCEPUTOBS= System Option" on page 365.

REDUCEPUTVALUES=n

when the REDUCEPUT= option is set to DBMS, BASE, or ALL, specifies the maximum number of SAS format values that can exist in a PUT function expression for PROC SQL to consider optimizing the PUT function in a query.

Default: 100

Range: 100 – 3,000

Requirement: *n* must be an integer

Interaction: If the number of SAS format values in a PUT function expression is greater than this value, PROC SQL does not optimize the PUT function.

Tips:

Alternatively, you can set the SQLREDUCEPUTVALUES= system option. The value that is specified in the SQLREDUCEPUTVALUES= system option is in effect for all SQL procedure statements, unless the REDUCEPUTVALUES= option is set. The value of the REDUCEPUTVALUES= option takes precedence over the SQLREDUCEPUTVALUES= system option. The RESET statement can also be used to set or reset the REDUCEPUTVALUES= option. However, changing the value of the REDUCEPUTVALUES= option does not change the value of the SQLREDUCEPUTVALUES= system option. For more information, see "SQLREDUCEPUTVALUES= System Option" on page 366.

The value for REDUCEPUTVALUES= is used for each individual optimization. For example, if you have a PUT function in a WHERE clause, and another PUT function in a SELECT statement, and both have user-defined formats with contained values, the value of REDUCEPUTVALUES= is applied separately for the clause and the statement.

REMERGE|NOREMERGE

Specifies whether PROC SQL can process queries that use remerging of data. The remerge feature of PROC SQL makes two passes through a table, using data in the second pass that was created in the first pass, in order to complete a query. When the NOREMERGE system option is set, PROC SQL cannot process remerging of data. If remerging is attempted when the NOREMERGE option is set, an error is written to the SAS log.

Default: REMERGE

Tip: Alternatively, you can set the SQLREMERGE system option. The value that is specified in the SQLREMERGE system option is in effect for all SQL procedure statements, unless the PROC SQL REMERGE option is set. The value of the REMERGE option takes precedence over the SQLREMERGE system option. The RESET statement can also be used to set or reset the REMERGE option. However, changing the value of the REMERGE option does not change the value of the SQLREMERGE system option. For more information, see "SQLREMERGE System Option" on page 368.

See: "Remerging Data" on page 350

SORTMSG|NOSORTMSG

Certain operations, such as ORDER BY, can sort tables internally using PROC SORT. Specifying SORTMSG requests information from PROC SORT about the sort and displays the information in the log.

Default: NOSORTMSG

SORTSEQ=*sort-table*

specifies the collating sequence to use when a query contains an ORDER BY clause. Use this option only if you want a collating sequence other than your system's or installation's default collating sequence.

See: SORTSEQ= option in *SAS National Language Support (NLS): Reference Guide*.

STIMER|NOSTIMER

specifies whether PROC SQL writes timing information to the SAS log for each statement, rather than as a cumulative value for the entire procedure. For this option to work, you must also specify the SAS system option STIMER. Some operating environments require that you specify this system option when you invoke SAS. If you use the system option alone, then you receive timing information for the entire SQL procedure, not on a statement-by-statement basis.

Default: NOSTIMER

STOPONTRUNC

specifies to not insert or update a row that contains data larger than the column when a truncation error occurs. This applies only when using the SET clause in an INSERT or UPDATE statement.

THREADS|NOTHREADS

overrides the SAS system option THREADS|NOTHREADS for a particular invocation of PROC SQL unless the system option is restricted. (See Restriction.) THREADS|NOTHREADS can also be specified in a RESET statement for use in particular queries. When THREADS is specified, PROC SQL uses parallel processing in order to increase the performance of sorting operations that involve large amounts of data. For more information about parallel processing, see *SAS Language Reference: Concepts*.

Default: value of SAS system option THREADS|NOTHREADS.

Restriction: Your site administrator can create a restricted options table. A restricted options table specifies SAS system option values that are established at start-up and cannot be overridden. If the THREADS | NOTHREADS system option is listed in the restricted options table, any attempt to set it is ignored and a warning message is written to the SAS log.

Interaction: When THREADS|NOTHREADS has been specified in a PROC SQL statement or a RESET statement, there is no way to reset the option to its default (that is, the value of the SAS system option THREADS|NOTHREADS) for that invocation of PROC SQL.

UNDO_POLICY=NONE|OPTIONAL|REQUIRED

specifies how PROC SQL handles updated data if errors occur while you are updating data. You can use UNDO_POLICY= to control whether your changes are permanent.

NONE

keeps any updates or inserts.

OPTIONAL

reverses any updates or inserts that it can reverse reliably.

REQUIRED

reverses all inserts or updates that have been done to the point of the error. In some cases, the UNDO operation cannot be done reliably. For example, when a program uses a SAS/ACCESS view, it might not be able to reverse the effects of the INSERT and UPDATE statements without reversing the effects of other changes at the same time. In that case, PROC SQL issues an error message and does not execute the statement. Also, when a SAS data set is accessed through a SAS/SHARE server and is opened with the data set option CNTLLEV=RECORD, you cannot reliably reverse your changes.

This option can enable other users to update newly inserted rows. If an error occurs during the insert, then PROC SQL can delete a record that another user updated. In that case, the statement is not executed, and an error message is issued.

Default: REQUIRED

Tips:

If you are updating a data set using the SPD Engine, you can significantly improve processing performance by setting UNDO_POLICY=NONE. However, ensure that NONE is an appropriate setting for your application.

Alternatively, you can set the SQLUNDOPOLICY system option. The value that is specified in the SQLUNDOPOLICY= system option is in effect for all SQL procedure statements, unless the PROC SQL UNDO_POLICY= option is set. The value of the UNDO_POLICY= option takes precedence over the SQLUNDOPOLICY= system option. The RESET statement can also be used to set or reset the UNDO_POLICY= option. However, changing the value of the UNDO_POLICY= option does not change the value of the SQLUNDOPOLICY= system option. After the procedure completes, it reverts to the value of the SQLUNDOPOLICY= system option. For more information, see the "SQLUNDOPOLICY= System Option" on page 368.

WARNRECURS|NOWARNRECURS

specifies whether a warning displays in the SAS log for recursive references.

NOWARNRECURS specifies to display recursive references in a note, instead of as a warning in the SAS log.

Default: WARNRECURS

Details

Note: Options can be added, removed, or changed between PROC SQL statements with the RESET Statement on page 242.

ALTER TABLE Statement

Adds columns to, drops columns from, and changes column attributes in an existing table. Adds, modifies, and drops integrity constraints from an existing table.

Restrictions: You cannot use any type of view in an ALTER TABLE statement.

You cannot use ALTER TABLE on a table that is accessed by an engine that does not support UPDATE processing.

You must use at least one ADD, DROP, or MODIFY clause in the ALTER TABLE statement.

See: "Example 3: Updating Data in a PROC SQL Table" on page 249

Syntax

ALTER TABLE *table-name*

<ADD CONSTRAINT *constraint-nameconstraint-clause*<, … *constraint-nameconstraint-clause*

<ADD*constraint-specification*<, … *constraint-specification*>>

<ADD column-definition<, … column-definition>>

<DROP CONSTRAINT*constraint-name*<, … *constraint-name*>>

<DROP*column*<, … *column*>>

<DROP FOREIGN KEY*constraint-name*>

<DROP PRIMARY KEY>

<MODIFY column-definition<, … column-definition>>

;

Required Arguments

<ADD CONSTRAINT *constraint-nameconstraint-specification*<, ... *constraint-nameconstraint-specification*>>

adds the integrity constraint that is specified in *constraint-specification* and assigns *constraint-name* to it.

<ADD *constraint-specification*<, ... *constraint-specification*>>

adds the integrity constraint that is specified in *constraint-specification* and assigns a default name to it. The default constraint name has the form that is shown in the following table:

Default Name	Constraint Type
NM*xxxx*	Not null
UN*xxxx*	Unique
CK*xxxx*	Check

PK*xxxx*	Primary key
FK*xxxx*	Foreign key

In these default names, *xxxx* is a counter that begins at 0001.

<ADD column-definition<, ... column-definition>>
adds the column or columns that are specified in each column-definition.

column
names a column in *table-name*.

column-definition
See "column-definition" on page 308

constraint
is one of the following integrity constraints:

CHECK *(WHERE-clause)*
specifies that all rows in *table-name* satisfy the *WHERE-clause*.

DISTINCT (*column*<, ... *column*>)
specifies that the values of each *column* must be unique. This constraint is identical to UNIQUE.

FOREIGN KEY (*column*<, ... *column*>)REFERENCES *table-name* <ON DELETE *referential-action*> <ON UPDATE *referential-action*>
specifies a foreign key, that is, a set of *columns* whose values are linked to the values of the primary key variable in another table (the *table-name* that is specified for REFERENCES). The *referential-actions* are performed when the values of a primary key column that is referenced by the foreign key are updated or deleted.

Restriction: When defining overlapping primary key and foreign key constraints, the variables in a data file are part of both a primary key and a foreign key definition. If you use the exact same variables, then the variables must be defined in a different order. The foreign key's update and delete referential actions must both be RESTRICT.

NOT NULL (*column*)
specifies that *column* does not contain a null or missing value, including special missing values.

PRIMARY KEY (*column*<,...*column*>)
specifies one or more primary key columns, that is, columns that do not contain missing values and whose values are unique.

Restriction: When you are defining overlapping primary key and foreign key constraints, the variables in a data file are part of both a primary key definition and a foreign key definition. If you use the exact same variables, then the variables must be defined in a different order.

UNIQUE (*column*<,...*column*>)
specifies that the values of each *column* must be unique. This constraint is identical to DISTINCT.

constraint-name
specifies a name for the constraint that is being specified. The name must be a valid SAS name.

Note: The names PRIMARY, FOREIGN, MESSAGE, UNIQUE, DISTINCT, CHECK, and NOT cannot be used as values for *constraint-name*.

constraint-specification
consists of

constraint <MESSAGE='*message-string*' <MSGTYPE=*message-type*>>

<DROP *column*<,...*column*>>
deletes each *column* from the table.

<DROP CONSTRAINT*constraint-name*<,...*constraint-name*>>
deletes the integrity constraint that is referenced by each *constraint-name*. To find the name of an integrity constraint, use the DESCRIBE TABLE CONSTRAINTS clause (see DESCRIBE Statement on page 237).

<DROP FOREIGN KEY *constraint-name*>
Removes the foreign key constraint that is referenced by *constraint-name*.

Note: The DROP FOREIGN KEY clause is a DB2 extension.

<DROP PRIMARY KEY>
Removes the primary key constraint from *table-name*.

Note: The DROP PRIMARY KEY clause is a DB2 extension.

message-string
specifies the text of an error message that is written to the log when the integrity constraint is not met. The maximum length of *message-string* is 250 characters.

message-type
specifies how the error message is displayed in the SAS log when an integrity constraint is not met.

NEWLINE
the text that is specified for MESSAGE= is displayed as well as the default error message for that integrity constraint.

USER
only the text that is specified for MESSAGE= is displayed.

<MODIFY column-definition<, ... column-definition>>
changes one or more attributes of the column that is specified in each column-definition.

referential-action
specifies the type of action to be performed on all matching foreign key values.

CASCADE
allows primary key data values to be updated, and updates matching values in the foreign key to the same values. This referential action is currently supported for updates only.

RESTRICT
prevents the update or deletion of primary key data values if a matching value exists in the foreign key. This referential action is the default.

SET NULL
allows primary key data values to be updated, and sets all matching foreign key values to NULL.

table-name

- in the ALTER TABLE statement, refers to the name of the table that is to be altered.

- in the REFERENCES clause, refers to the name of table that contains the primary key that is referenced by the foreign key.

table-name can be a one-level name, a two-level *libref.table* name, or a physical pathname that is enclosed in single quotation marks.

WHERE-clause
specifies a SAS WHERE clause. Do not include the WHERE keyword in the WHERE clause.

Details

Specifying Initial Values of New Columns

When the ALTER TABLE statement adds a column to the table, it initializes the column's values to missing in all rows of the table. Use the UPDATE statement to add values to the new column or columns.

Changing Column Attributes

If a column is already in the table, then you can change the following column attributes by using the MODIFY clause: length, informat, format, and label. The values in a table are either truncated or padded with blanks (if character data) as necessary to meet the specified length attribute.

You cannot change a character column to numeric and vice versa. To change a column's data type, drop the column and then add it (and its data) again, or use the DATA step.

Note: You cannot change the length of a numeric column with the ALTER TABLE statement. Use the DATA step instead.

Renaming Columns

You cannot use the RENAME= data set option with the ALTER TABLE statement to change a column's name. However, you can use the RENAME= data set option with the CREATE TABLE or SELECT statement. For more information about the RENAME= data set option, see the section on SAS data set options in *SAS Data Set Options: Reference*.

Indexes on Altered Columns

When you alter the attributes of a column and an index has been defined for that column, the values in the altered column continue to have the index defined for them. If you drop a column with the ALTER TABLE statement, then all the indexes (simple and composite) in which the column participates are also dropped. See CREATE INDEX Statement on page 228 for more information about creating and using indexes.

Integrity Constraints

Use ALTER TABLE to modify integrity constraints for existing tables. Use the CREATE TABLE statement to attach integrity constraints to new tables. For more information about integrity constraints, see the section on SAS files in *SAS Language Reference: Concepts*.

CONNECT Statement

Establishes a connection with a DBMS that SAS/ACCESS software supports.

Requirement: SAS/ACCESS software is required. For more information about this statement, see your SAS/ACCESS documentation.

See: “Connecting to a DBMS by Using the SQL Procedure Pass-Through Facility” on page 166

Syntax

CONNECT TO *dbms-name* <AS *alias*>
<(*connect-statement-argument-1*=*value*<… *connect-statement-argument-n*=*value*>)>
<(*database-connection-argument-1*=*value*<… *database-connection-argument-n*=*value*>)>;

CONNECT USING *libname* <AS *alias*>;

Required Arguments

alias
specifies an alias that has 1 to 32 characters. The keyword AS must precede *alias*. Some DBMSs allow more than one connection. The optional AS clause enables you to name the connections so that you can refer to them later.

connect-statement-argument=value
specifies values for arguments that indicate whether you can make multiple connections, shared or unique connections, and so on, to the database. These arguments are optional, but if they are included, then they must be enclosed in parentheses. See *SAS/ACCESS for Relational Databases: Reference* for more information about these arguments.

database-connection-argument=value
specifies values for the DBMS-specific arguments that are needed by PROC SQL in order to connect to the DBMS. These arguments are optional for most databases, but if they are included, then they must be enclosed in parentheses. For more information, see the SAS/ACCESS documentation for your DBMS.

dbms-name
identifies the DBMS that you want to connect to (for example, ORACLE or DB2).

libname
specifies the LIBNAME where a DBMS connection has already been established. The LIBNAME can be reused in the SQL procedure using the CONNECT statement.

CREATE INDEX Statement

Creates indexes on columns in tables.

Restriction: You cannot use CREATE INDEX on a table that is accessed with an engine that does not support UPDATE processing.

Syntax

CREATE <UNIQUE> **INDEX** *index-name*
ON *table-name* (*column* <, … *column*>);

Required Arguments

column
specifies a column in *table-name*.

index-name
names the index that you are creating. If you are creating an index on one column only, then *index-name* must be the same as *column*. If you are creating an index on more than one column, then *index-name* cannot be the same as any column in the table.

table-name
specifies a PROC SQL table.

Details

Indexes in PROC SQL

An index stores both the values of a table's columns and a system of directions that enable access to rows in that table by index value. Defining an index on a column or set of columns enables SAS, under certain circumstances, to locate rows in a table more quickly and efficiently. Indexes enable PROC SQL to execute the following classes of queries more efficiently:

- comparisons against a column that is indexed
- an IN subquery where the column in the inner subquery is indexed
- correlated subqueries, where the column being compared with the correlated reference is indexed
- join-queries, where the join-expression is an equals comparison and all the columns in the join-expression are indexed in one of the tables being joined.

SAS maintains indexes for all changes to the table, whether the changes originate from PROC SQL or from some other source. Therefore, if you alter a column's definition or update its values, then the same index continues to be defined for it. However, if an indexed column in a table is dropped, then the index on it is also dropped.

You can create simple or composite indexes. A *simple index* is created on one column in a table. A simple index must have the same name as that column. A *composite index* is one index name that is defined for two or more columns. The columns can be specified in any order, and they can have different data types. A composite index name cannot match the name of any column in the table. If you drop a composite index, then the index is dropped for all the columns named in that composite index.

UNIQUE Keyword

The UNIQUE keyword causes SAS to reject any change to a table that would cause more than one row to have the same index value. Unique indexes guarantee that data in one column, or in a composite group of columns, remains unique for every row in a table. A unique index can be defined for a column that includes NULL or missing values if each row has a unique index value.

Managing Indexes

You can use the CONTENTS statement in the DATASETS procedure to display a table's index names and the columns for which they are defined. You can also use the DICTIONARY tables INDEXES, TABLES, and COLUMNS to list information about indexes. For more information, see "Accessing SAS System Information by Using DICTIONARY Tables" on page 144.

See the section on SAS files in *SAS Language Reference: Concepts* for a further description of when to use indexes and how they affect SAS statements that handle BY-group processing.

CREATE TABLE Statement

Creates PROC SQL tables.

See: "Example 1: Creating a Table and Inserting Data into It" on page 245
"Example 2: Creating a Table from a Query's Result" on page 247

Syntax

CREATE TABLE *table-name*
(*column-specification*<, ...*column-specification* | *constraint-specification*>)
;

CREATE TABLE *table-name* **LIKE** *table-name2*;

CREATE TABLE *table-name* **AS** query-expression
<ORDER BY*order-by-item*<, ... *order-by-item*>>;

Required Arguments

column-constraint
is one of the following:

CHECK (*WHERE-clause*
specifies that all rows in *table-name* satisfy the *WHERE-clause*.

DISTINCT
specifies that the values of the column must be unique. This constraint is identical to UNIQUE.

NOT NULL
specifies that the column does not contain a null or missing value, including special missing values.

PRIMARY KEY
specifies that the column is a primary key column, that is, a column that does not contain missing values and whose values are unique.

Restriction: When defining overlapping primary key and foreign key constraints, the variables in a data file are part of both a primary key and a foreign key definition. If you use the exact same variables, then the variables must be defined in a different order.

REFERENCES*table-name* <ON DELETE *referential-action*><ON UPDATE *referential-action*>
specifies that the column is a foreign key, that is, a column whose values are linked to the values of the primary key variable in another table (the *table-name* that is specified for REFERENCES). The *referential-actions* are performed when the values of a primary key column that is referenced by the foreign key are updated or deleted.

Restriction: When you are defining overlapping primary key and foreign key constraints, the variables in a data file are part of both a primary key definition and a foreign key definition. If you use the exact same variables, then the variables must be defined in a different order. The foreign key's update and delete referential actions must both be RESTRICT.

UNIQUE
specifies that the values of the column must be unique. This constraint is identical to DISTINCT.

Note: If you specify *column-constraint*, then SAS automatically assigns a name to the constraint. The constraint name has the form the following form, where *xxxx* is a counter that begins at 0001.

Default name	Constraint type
CK*xxxx*	Check
FK*xxxx*	Foreign key
NM*xxxx*	Not Null
PK*xxxx*	Primary key
UN*xxxx*	Unique

column-definition
See "column-definition" on page 308

column-specification
consists of

column-definition <*column-constraint*>

constraint
is one of the following:

CHECK *WHERE-clause*
specifies that all rows in *table-name* satisfy the *WHERE-clause*.

DISTINCT (*column*<, ... *column*>)
specifies that the values of each *column* must be unique. This constraint is identical to UNIQUE.

FOREIGN KEY (*column*<, ... *column*>)
REFERENCES *table-name*<ON DELETE *referential-action*> <ON UPDATE *referential-action*>
specifies a foreign key, that is, a set of *columns* whose values are linked to the values of the primary key variable in another table (the *table-name* that is specified for REFERENCES). The *referential-actions* are performed when the values of a primary key column that is referenced by the foreign key are updated or deleted.

Restriction: When you are defining overlapping primary key and foreign key constraints, the variables in a data file are part of both a primary key definition and a foreign key definition. If you use the exact same variables, then the variables must be defined in a different order. The foreign key's update and delete referential actions must both be RESTRICT.

NOT NULL (*column*)
specifies that *column* does not contain a null or missing value, including special missing values.

PRIMARY KEY (*column*<, ...column>)
specifies one or more primary key columns, that is, columns that do not contain missing values and whose values are unique.

Restriction: When defining overlapping primary key and foreign key constraints, the variables in a data file are part of both a primary key and a foreign key definition. If you use the exact same variables, then the variables must be defined in a different order.

UNIQUE (*column*<, ...column>)
specifies that the values of each *column* must be unique. This constraint is identical to DISTINCT.

constraint-name
specifies a name for the constraint that is being specified. The name must be a valid SAS name.

Note: The names PRIMARY, FOREIGN, MESSAGE, UNIQUE, DISTINCT, CHECK, and NOT cannot be used as values for *constraint-name*.

constraint-specification
consists of

CONSTRAINT *constraint-name constraint* <MESSAGE='*message-string*' <MSGTYPE=*message-type*>>

message-string
specifies the text of an error message that is written to the log when the integrity constraint is not met. The maximum length of *message-string* is 250 characters.

message-type
specifies how the error message is displayed in the SAS log when an integrity constraint is not met.

NEWLINE
the text that is specified for MESSAGE= is displayed as well as the default error message for that integrity constraint.

USER
only the text that is specified for MESSAGE= is displayed.

ORDER BY *order-by-item*
sorts the rows in *table-name* by the values of each *order-by-item*. See “ORDER BY Clause” on page 301.

query-expression
creates *table-name* from the results of a query. See “query-expression” on page 330.

referential-action
specifies the type of action to be performed on all matching foreign key values.

CASCADE
allows primary key data values to be updated, and updates matching values in the foreign key to the same values. This referential action is currently supported for updates only.

RESTRICT
occurs only if there are matching foreign key values. This referential action is the default.

SET NULL
sets all matching foreign key values to NULL.

table-name

- in the CREATE TABLE statement, refers to the name of the table that is to be created. You can use data set options by placing them in parentheses immediately after *table-name*. See "Using SAS Data Set Options with PROC SQL" on page 151.
- in the REFERENCES clause, refers to the name of table that contains the primary key that is referenced by the foreign key.

table-name2
creates *table-name* with the same column names and column attributes as *table-name2*, but with no rows.

WHERE-clause
specifies a SAS WHERE clause. Do not include the WHERE keyword in the WHERE clause.

Details

Creating a Table without Rows

- The first form of the CREATE TABLE statement creates tables that automatically map SQL data types to tables that are supported by SAS. Use this form when you want to create a new table with columns that are not present in existing tables. It is also useful if you are running SQL statements from an SQL application in another SQL-based database.
- The second form uses a LIKE clause to create a table that has the same column names and column attributes as another table. To drop any columns in the new table, you can specify the DROP= data set option in the CREATE TABLE statement. The specified columns are dropped when the table is created. Indexes are not copied to the new table.

  Both of these forms create a table without rows. You can use an INSERT statement to add rows. Use an ALTER TABLE statement to modify column attributes or to add or drop columns.

Creating a Table from a Query Expression

- The third form of the CREATE TABLE statement stores the results of any query expression in a table and does not display the output. It is a convenient way to create temporary tables that are subsets or supersets of other tables.

  When you use this form, a table is physically created as the statement is executed. The newly created table does not reflect subsequent changes in the underlying tables (in the query expression). If you want to continually access the most current data, then create a view from the query expression instead of a table. See CREATE VIEW Statement on page 234.

  CAUTION:

  Recursive table references can cause data integrity problems. While it is possible to recursively reference the target table of a CREATE TABLE AS statement, doing so can cause data integrity problems and incorrect results. Constructions such as the following should be avoided: `proc sql; create table a as select var1, var2 from a;`

Integrity Constraints

You can attach integrity constraints when you create a new table. To modify integrity constraints, use the ALTER TABLE statement.

The following interactions apply to integrity constraints when they are part of a column specification.

- You cannot specify compound primary keys.
- The check constraint that you specify in a column specification does not need to reference that same column in its WHERE clause.
- You can specify more than one integrity constraint.
- You can specify the MSGTYPE= and MESSAGE= options on a constraint.

For more information about integrity constraints, see the section on SAS files in *SAS Language Reference: Concepts*.

CREATE VIEW Statement

Creates a PROC SQL view from a query expression.

See: “What Are Views?” on page 211
“Example 8: Creating a View from a Query's Result” on page 265

Syntax

CREATE VIEW *proc-sql-view* <(*column-name-list*)> **AS** query-expression
<ORDER BY*order-by-item*<, ... *order-by-item*>>
<USING*libname-clause*<, ... *libname-clause*>> ;

Required Arguments

column-name-list
is a comma-separated list of column names for the view, to be used in place of the column names or aliases that are specified in the SELECT clause. The names in this list are assigned to columns in the order in which they are specified in the SELECT clause. If the number of column names in this list does not equal the number of columns in the SELECT clause, then a warning is written to the SAS log.

query-expression
See “query-expression” on page 330

libname-clause
is one of the following:

LIBNAME *libref* <*engine*> *'SAS-library'* <*option(s)*> <*engine-host-option(s)*>

LIBNAME *libref SAS/ACCESS-engine-name*<*SAS/ACCESS-engine-connection-option(s)*> <*SAS/ACCESS-engine-LIBNAME-option(s)*>
See *SAS Statements: Reference* for information about the Base SAS LIBNAME statement. See *SAS/ACCESS for Relational Databases: Reference* for information about the LIBNAME statement for relational databases.

order-by-item
See “ORDER BY Clause” on page 301

proc-sql-view
specifies the name for the PROC SQL view that you are creating. See "What Are Views?" on page 211 for a definition of a PROC SQL view.

Details

Sorting Data Retrieved by Views

PROC SQL enables you to specify the ORDER BY clause in the CREATE VIEW statement. When a view with an ORDER BY clause is accessed, and the ORDER BY clause directly affects the order of the results, its data is sorted and displayed as specified by the ORDER BY clause. However, if the ORDER BY clause does not directly affect the order of the results (for example, if the view is specified as part of a join), then PROC SQL ignores the ORDER BY clause in order to enhance performance.

Note: If the ORDER BY clause is omitted, then a particular order to the output rows, such as the order in which the rows are encountered in the queried table, cannot be guaranteed—even if an index is present. Without an ORDER BY clause, the order of the output rows is determined by the internal processing of PROC SQL, the default collating sequence of SAS, and your operating environment. Therefore, if you want your results to appear in a particular order, then use the ORDER BY clause.

Note: If you specify the NUMBER option in the PROC SQL statement when you create your view, then the ROW column appears in the output. However, you cannot order by the ROW column in subsequent queries. See the description of NUMBER| NONUMBER on page 220.

Librefs and Stored Views

You can refer to a table name alone (without the libref) in the FROM clause of a CREATE VIEW statement if the table and view reside in the same SAS library, as in this example:

```
create view proclib.view1 as
   select *
      from invoice
      where invqty>10;
```

In this view, VIEW1 and INVOICE are stored permanently in the SAS library referenced by PROCLIB. Specifying a libref for INVOICE is optional.

Updating Views

You can update a view's underlying data with some restrictions. See "Updating PROC SQL and SAS/ACCESS Views" on page 168.

Embedded LIBNAME Statements

The USING clause enables you to store DBMS connection information in a view by embedding the SAS/ACCESS LIBNAME statement inside the view. When PROC SQL executes the view, the stored query assigns the libref and establishes the DBMS connection using the information in the LIBNAME statement. The scope of the libref is local to the view, and will not conflict with any identically named librefs in the SAS session. When the query finishes, the connection to the DBMS is terminated and the libref is deassigned.

The USING clause must be the last clause in the CREATE VIEW statement. Multiple LIBNAME statements can be specified, separated by commas. In the following example, a connection is made and the libref ACCREC is assigned to an ORACLE database.

```
create view proclib.view1 as
   select *
      from accrec.invoices as invoices
      using libname accrec oracle
         user=username
pass=password
         path='dbms-path';
```

For more information about the SAS/ACCESS LIBNAME statement, see the SAS/ACCESS documentation for your DBMS.

Note: Starting in SAS System 9, PROC SQL views, the pass-through facility, and the SAS/ACCESS LIBNAME statement are the preferred ways to access relational DBMS data; SAS/ACCESS views are no longer recommended. You can convert existing SAS/ACCESS views to PROC SQL views by using the CV2VIEW procedure. For more information, see Chapter 32, "CV2VIEW Procedure" in *SAS/ACCESS for Relational Databases: Reference*.

You can also embed a SAS LIBNAME statement in a view with the USING clause, which enables you to store SAS libref information in the view. Just as in the embedded SAS/ACCESS LIBNAME statement, the scope of the libref is local to the view, and it will not conflict with an identically named libref in the SAS session.

```
create view work.tableview as
   select * from proclib.invoices
      using libname proclib
'SAS-library';
```

DELETE Statement

Removes one or more rows from a table or view that is specified in the FROM clause.

Restriction: You cannot use DELETE FROM on a table that is accessed by an engine that does not support UPDATE processing.

See: "Example 5: Combining Two Tables" on page 254

Syntax

DELETE
FROM *table-name|sas/access-view|proc-sql-view* <AS*alias*>
<WHERE sql-expression>;

Required Arguments

alias
assigns an alias to *table-name*, *sas/access-view*, or *proc-sql-view*.

sas/access-view
specifies a SAS/ACCESS view that you are deleting rows from.

proc-sql-view
specifies a PROC SQL view that you are deleting rows from. *proc-sql-view* can be a one-level name, a two-level *libref.view* name, or a physical pathname that is enclosed in single quotation marks.

sql-expression
See "sql-expression" on page 338

table-name
specifies the table that you are deleting rows from. *table-name* can be a one-level name, a two-level *libref.table* name, or a physical pathname that is enclosed in single quotation marks.

CAUTION:
Recursive table references can cause data integrity problems. While it is possible to recursively reference the target table of a DELETE statement, doing so can cause data integrity problems and incorrect results. Constructions such as the following should be avoided:

```
proc sql;
   delete from a
      where var1 > (select min(var2) from a);
```

Details

Deleting Rows through Views

You can delete one or more rows from a view's underlying table, with some restrictions. See "Updating PROC SQL and SAS/ACCESS Views" on page 168.

CAUTION:
If you omit a WHERE clause, the DELETE statement deletes all of the rows from the specified table or the table that is described by a view. The rows are not actually deleted from the table until it is recreated.

DESCRIBE Statement

Displays a PROC SQL definition in the SAS log.

Restriction: PROC SQL views are the only type of view allowed in a DESCRIBE VIEW statement.

See: "Example 6: Reporting from DICTIONARY Tables" on page 257

Syntax

DESCRIBE TABLE *table-name* <, … *table-name*>;

DESCRIBE VIEW *proc-sql-view* <, … *proc-sql-view*>;

DESCRIBE TABLE CONSTRAINTS *table-name* <, … *table-name*>;

Required Arguments

table-name
specifies a PROC SQL table. *table-name* can be a one-level name, a two-level *libref.table* name, or a physical pathname that is enclosed in single quotation marks.

proc-sql-view
specifies a PROC SQL view. *proc-sql-view* can be a one-level name, a two-level *libref.view* name, or a physical pathname that is enclosed in single quotation marks.

Details

- The DESCRIBE TABLE statement writes a CREATE TABLE statement to the SAS log for the table specified in the DESCRIBE TABLE statement, regardless of how the table was originally created (for example, with a DATA step). If applicable, SAS data set options are included with the table definition. If indexes are defined on columns in the table, then CREATE INDEX statements for those indexes are also written to the SAS log.

When you are transferring a table to a DBMS that SAS/ACCESS software supports, it is helpful to know how it is defined. To find out more information about a table, use the FEEDBACK option or the CONTENTS statement in the DATASETS procedure.

- The DESCRIBE VIEW statement writes a view definition to the SAS log. If you use a PROC SQL view in the DESCRIBE VIEW statement that is based on or derived from another view, then you might want to use the FEEDBACK option in the PROC SQL statement. This option displays in the SAS log how the underlying view is defined and expands any expressions that are used in this view definition. The CONTENTS statement in DATASETS procedure can also be used with a view to find out more information.
- The DESCRIBE TABLE CONSTRAINTS statement lists the integrity constraints that are defined for the specified table or tables. However, names of the foreign key data set variables that reference the primary key constraint will not be displayed as part of the primary key constraint's DESCRIBE TABLE output.

DISCONNECT Statement

Ends the connection with a DBMS that a SAS/ACCESS interface supports.

Requirement: SAS/ACCESS software is required. For more information about this statement, see your SAS/ACCESS documentation.

See: "Connecting to a DBMS by Using the SQL Procedure Pass-Through Facility" on page 166

Syntax

DISCONNECT FROM *dbms-name|alias*;

Required Arguments

alias
specifies the alias that is defined in the CONNECT statement.

dbms-name
specifies the DBMS from which you want to end the connection (for example, DB2 or ORACLE). The name that you specify should match the name that is specified in the CONNECT statement.

Details

- An implicit COMMIT is performed before the DISCONNECT statement ends the DBMS connection. If a DISCONNECT statement is not submitted, then implicit DISCONNECT and COMMIT actions are performed and the connection to the DBMS is broken when PROC SQL terminates.
- PROC SQL continues executing until you submit a QUIT statement, another SAS procedure, or a DATA step.

DROP Statement

Deletes tables, views, or indexes.

Restriction: You cannot use DROP TABLE or DROP INDEX on a table that is accessed by an engine that does not support UPDATE processing.

Syntax

DROP TABLE *table-name* <, ... *table-name*>;

DROP VIEW *view-name* <, ... *view-name*>;

DROP INDEX *index-name* <, ... *index-name*>
FROM *table-name*;

Required Arguments

index-name
specifies an index that exists on *table-name*.

table-name
specifies a PROC SQL table. *table-name* can be a one-level name, a two-level *libref.table* name, or a physical pathname that is enclosed in single quotation marks.

view-name
specifies a SAS view of any type: PROC SQL view, SAS/ACCESS view, or DATA step view. *view-name* can be a one-level name, a two-level *libref.view* name, or a physical pathname that is enclosed in single quotation marks.

Details

- If you drop a table that is referenced in a view definition and try to execute the view, then an error message is written to the SAS log that states that the table does not exist. Therefore, remove references in queries and views to any tables and views that you drop.
- If you drop a table with indexed columns, then all the indexes are automatically dropped. If you drop a composite index, then the index is dropped for all the columns that are named in that index.
- You can use the DROP statement to drop a table or view in an external database that is accessed with the pass-through facility or SAS/ACCESS LIBNAME statement, but not for an external database table or view that a SAS/ACCESS view describes.

EXECUTE Statement

Sends a DBMS-specific SQL statement to a DBMS that a SAS/ACCESS interface supports.

Requirement: SAS/ACCESS software is required. For more information about this statement, see your SAS/ACCESS documentation.

See: "Connecting to a DBMS by Using the SQL Procedure Pass-Through Facility" on page 166

SQL documentation for your DBMS

Syntax

EXECUTE (*dbms-SQL-statement*)
BY *dbms-name|alias*;

Required Arguments

alias
specifies an optional alias that is defined in the CONNECT statement. Note that *alias* must be preceded by the keyword BY.

dbms-name
identifies the DBMS to which you want to direct the DBMS statement (for example, ORACLE or DB2).

dbms-SQL-statement
is any DBMS-specific SQL statement, except the SELECT statement, which can be executed by the DBMS-specific dynamic SQL. The SQL statement can contain a semicolon. The SQL statement can be case-sensitive, depending on your data source, and it is passed to the data source exactly as you type it.

Details

- If your DBMS supports multiple connections, then you can use the alias that is defined in the CONNECT statement. This alias directs the EXECUTE statements to a specific DBMS connection.
- Any return code or message that is generated by the DBMS is available in the macro variables SQLXRC and SQLXMSG after the statement completes.

Example

The following example, after the connection, uses the EXECUTE statement to drop a table, create a table, and insert a row of data.

```
proc sql;
    execute(drop table ' My Invoice ') by db;
    execute(create table ' My Invoice '(
       ' Invoice Number ' LONG not null,
       ' Billed To '  VARCHAR(20),
       ' Amount '  CURRENCY,
       ' BILLED ON '  DATETIME)) by db;
    execute(insert into  ' My Invoice '
       values( 12345, 'John Doe', 123.45, #11/22/2003#)) by db;
quit;
```

INSERT Statement

Adds rows to a new or existing table or view.

Restriction: You cannot use INSERT INTO on a table that is accessed with an engine that does not support UPDATE processing.

See: "Example 1: Creating a Table and Inserting Data into It" on page 245

Syntax

INSERT INTO *table-name|sas/access-view|proc-sql-view* <(*column*<, ... *column*>)>

SET *column*=sql-expression
<, ... *column*=sql-expression>

<SET*column*=sql-expression <, ... *column*=sql-expression>>;

INSERT INTO *table-name|sas/access-view|proc-sql-view* <(*column*<, ... *column*>)>

VALUES (*value* <, ... *value*>)
<... VALUES (*value*<, ... *value*>)>;

INSERT INTO *table-name|sas/access-view|proc-sql-view*
<(*column*<, ...*column*>)> query-expression;

Required Arguments

column
specifics the column into which you are inserting rows.

proc-sql-view
specifies a PROC SQL view into which you are inserting rows. *proc-sql-view* can be a one-level name, a two-level *libref.view* name, or a physical pathname that is enclosed in single quotation marks.

query-expression
See "query-expression" on page 330

sas/access-view
specifies a SAS/ACCESS view into which you are inserting rows.

sql-expression
See "sql-expression" on page 338

Restriction: You cannot use a logical operator (AND, OR, or NOT) in an expression in a SET clause.

table-name
specifies a PROC SQL table into which you are inserting rows. *table-name* can be a one-level name, a two-level *libref.table* name, or a physical pathname that is enclosed in single quotation marks.

value
is a data value.

CAUTION:

Recursive table references can cause data integrity problems. While it is possible to recursively reference the target table of an INSERT statement, doing so can cause data integrity problems and incorrect results. Constructions such as the following should be avoided:

```
proc sql;
   insert into a
      select var1, var2
      from a
      where var1 > 0;
```

Details

Methods for Inserting Values

- The first form of the INSERT statement uses the SET clause, which specifies or alters the values of a column. You can use more than one SET clause per INSERT statement, and each SET clause can set the values in more than one column. Multiple SET clauses are not separated by commas. If you specify an optional list of columns, then you can set a value only for a column that is specified in the list of columns to be inserted.
- The second form of the INSERT statement uses the VALUES clause. This clause can be used to insert lists of values into a table. You can either give a value for each column in the table or give values just for the columns specified in the list of column names. One row is inserted for each VALUES clause. Multiple VALUES clauses are not separated by commas. The order of the values in the VALUES clause matches the order of the column names in the INSERT column list or, if no list was specified, the order of the columns in the table.
- The third form of the INSERT statement inserts the results of a query expression into a table. The order of the values in the query expression matches the order of the column names in the INSERT column list or, if no list was specified, the order of the columns in the table.

Note: If the INSERT statement includes an optional list of column names, then only those columns are given values by the statement. Columns that are in the table but not listed are given missing values.

Inserting Rows through Views

You can insert one or more rows into a table through a view, with some restrictions. See "Updating PROC SQL and SAS/ACCESS Views" on page 168.

Adding Values to an Indexed Column

If an index is defined on a column and you insert a new row into the table, then that value is added to the index. You can display information about indexes with

- the CONTENTS statement in the DATASETS procedure. See the XisError: User must supply overrideFetchedText because link target element has no title.
- the DICTIONARY.INDEXES table. For more information, see "Accessing SAS System Information by Using DICTIONARY Tables" on page 144.

For more information about creating and using indexes, see the CREATE INDEX Statement on page 228.

RESET Statement

Resets PROC SQL options without restarting the procedure.

See: "Example 5: Combining Two Tables" on page 254

Syntax

RESET *<option(s)>*;

Required Argument

RESET
The RESET statement enables you to add, drop, or change the options in PROC SQL without restarting the procedure. For more information, see PROC SQL Statement on page 215 for a description of the options.

SELECT Statement

Selects columns and rows of data from tables and views.

Restriction: The clauses in the SELECT statement must appear in the order shown.

See: Chapter 8, "SQL SELECT Statement Clauses," on page 289
"table-expression" on page 355
"query-expression" on page 330

Syntax

SELECT <DISTINCT> *object-item* <, *...object-item*>
<INTO *macro-variable-specification*<, ... *macro-variable-specification*>>
FROM *from-list*
<WHERE sql-expression>
<GROUP BY*group-by-item*<, ... *group-by-item*>>
<HAVING sql-expression>
<ORDER BY*order-by-item*<, ... *order-by-item*>>;

UPDATE Statement

Modifies a column's values in existing rows of a table or view.

Restriction: You cannot use UPDATE on a table that is accessed by an engine that does not support UPDATE processing.

See: "Example 3: Updating Data in a PROC SQL Table" on page 249

Syntax

UPDATE *table-name|sas/access-view|proc-sql-view* <AS*alias*>
SET *column*=sql-expression
<, ... *column*=sql-expression>
<SET*column*=sql-expression <, ... *column*=sql-expression>>
<WHERE sql-expression>;

Required Arguments

alias
assigns an alias to *table-name*, *sas/access-view*, or *proc-sql-view*.

column
specifies a column in *table-name*, *sas/access-view*, or *proc-sql-view*.

sas/access-view
specifies a SAS/ACCESS view.

sql-expression
See "sql-expression" on page 338

Restriction: You cannot use a logical operator (AND, OR, or NOT) in an expression in a SET clause.

table-name
specifies a PROC SQL table. *table-name* can be a one-level name, a two-level *libref.table* name, or a physical pathname that is enclosed in single quotation marks.

proc-sql-view
specifies a PROC SQL view. *proc-sql-view* can be a one-level name, a two-level *libref.view* name, or a physical pathname that is enclosed in single quotation marks.

Details

You can update one or more rows of a table through a view, with some restrictions. See "Updating PROC SQL and SAS/ACCESS Views" on page 168.

- Any column that is not modified retains its original values, except in certain queries using the CASE expression. See "CASE Expression" on page 306 for a description of CASE expressions.
- To add, drop, or modify a column's definition or attributes, use the ALTER TABLE statement, described in ALTER TABLE Statement on page 224.
- In the SET clause, a column reference on the left side of the equal sign can also appear as part of the expression on the right side of the equal sign. For example, you could use this expression to give employees a $1,000 holiday bonus:

  ```
  set salary=salary + 1000
  ```
- If you omit the WHERE clause, then all the rows are updated. When you use a WHERE clause, only the rows that meet the WHERE condition are updated.
- When you update a column and an index has been defined for that column, the values in the updated column continue to have the index defined for them.

VALIDATE Statement

Checks the accuracy of a query expression's syntax and semantics without executing the expression.

Syntax

VALIDATE query-expression;

Required Argument

query-expression
See "query-expression" on page 330

Details

- The VALIDATE statement writes a message in the SAS log that states that the query is valid. If there are errors, then VALIDATE writes error messages to the SAS log.

- The VALIDATE statement can also be included in applications that use the macro facility. When used in such an application, VALIDATE returns a value that indicates the query expression's validity. The value is returned through the macro variable SQLRC (a short form for SQL return code). For example, if a SELECT statement is valid, then the macro variable SQLRC returns a value of 0. See "Using the PROC SQL Automatic Macro Variables" on page 157 for more information.

Examples: SQL Procedure

Example 1: Creating a Table and Inserting Data into It

Features: CREATE TABLE statement
column-modifier
INSERT statement
VALUES clause
SELECT clause
FROM clause

Table name: PROCLIB.PAYLIST

This example creates the table PROCLIB.PAYLIST and inserts data into it.

Program

```
libname proclib 'SAS-library';
proc sql;
   create table proclib.paylist
       (IdNum char(4),
        Gender char(1),
        Jobcode char(3),
        Salary num,
        Birth num informat=date7.
                  format=date7.,
        Hired num informat=date7.
                  format=date7.);
insert into proclib.paylist
    values('1639','F','TA1',42260,'26JUN70'd,'28JAN91'd)
    values('1065','M','ME3',38090,'26JAN54'd,'07JAN92'd)
    values('1400','M','ME1',29769.'05NOV67'd,'16OCT90'd)
values('1561','M',null,36514,'30NOV63'd,'07OCT87'd)
    values('1221','F','FA3',.,'22SEP63'd,'04OCT94'd);
title 'PROCLIB.PAYLIST Table';
select *
   from proclib.paylist;
proc printto; run;
```

Program Description

Declare the PROCLIB library. The PROCLIB library is used in these examples to store created tables.

```
libname proclib 'SAS-library';
```

Create the PROCLIB.PAYLIST table. The CREATE TABLE statement creates PROCLIB.PAYLIST with six empty columns. Each column definition indicates whether the column is character or numeric. The number in parentheses specifies the width of the column. INFORMAT= and FORMAT= assign date informats and formats to the Birth and Hired columns.

```
proc sql;
   create table proclib.paylist
       (IdNum char(4),
        Gender char(1),
        Jobcode char(3),
        Salary num,
        Birth num informat=date7.
                  format=date7.,
        Hired num informat=date7.
                  format=date7.);
```

Insert values into the PROCLIB.PAYLIST table. The INSERT statement inserts data values into PROCLIB.PAYLIST according to the position in the VALUES clause. Therefore, in the first VALUES clause, *1639* is inserted into the first column, *F* into the second column, and so on. Dates in SAS are stored as integers with 0 equal to January 1, 1960. Suffixing the date with a *d* is one way to use the internal value for dates.

```
insert into proclib.paylist
    values('1639','F','TA1',42260,'26JUN70'd,'28JAN91'd)
    values('1065','M','ME3',38090,'26JAN54'd,'07JAN92'd)
    values('1400','M','ME1',29769.'05NOV67'd,'16OCT90'd)
```

Include missing values in the data. The value null represents a missing value for the character column Jobcode. The period represents a missing value for the numeric column Salary.

```
values('1561','M',null,36514,'30NOV63'd,'07OCT87'd)
    values('1221','F','FA3',.,'22SEP63'd,'04OCT94'd);
```

Specify the title.

```
title 'PROCLIB.PAYLIST Table';
```

Display the entire PROCLIB.PAYLIST table. The SELECT clause selects columns from PROCLIB.PAYLIST. The asterisk (*) selects all columns. The FROM clause specifies PROCLIB.PAYLIST as the table to select from.

```
select *
   from proclib.paylist;

proc printto; run;
```

HTML Output

Output 7.1 *Inserting Data into a Table*

PROCLIB.PAYLIST Table

IdNum	Gender	Jobcode	Salary	Birth	Hired
1639	F	TA1	42260	26JUN70	28JAN91
1065	M	ME3	38090	26JAN54	07JAN92
1400	M	ME1	29769	05NOV67	16OCT90
1561	M		36514	30NOV63	07OCT87
1221	F	FA3	.	22SEP63	04OCT94

Example 2: Creating a Table from a Query's Result

Features: CREATE TABLE statement
  AS query expression

SELECT clause
  column alias
  FORMAT= column-modifier
  object-item

Other features: data set option
  OBS=

Table name: PROCLIB.PAYROLL

Table name: PROCLIB.BONUS

Details

This example builds a column with an arithmetic expression and creates the PROCLIB.BONUS table from the query's result.

```
proc sql outobs=10;
   title 'PROCLIB.PAYROLL';
   title2 'First 10 Rows Only';
   select * from proclib.payroll;
   title;
```

***Figure 7.2** Query Result from PROCLIB.PAYROLL*

PROCLIB.PAYROLL
First 10 Rows Only

IdNumber	Gender	Jobcode	Salary	Birth	Hired
1919	M	TA2	34376	12SEP60	04JUN87
1653	F	ME2	35108	15OCT64	09AUG90
1400	M	ME1	29769	05NOV67	16OCT90
1350	F	FA3	32886	31AUG65	29JUL90
1401	M	TA3	38822	13DEC50	17NOV85
1499	M	ME3	43025	26APR54	07JUN80
1101	M	SCP	18723	06JUN62	01OCT90
1333	M	PT2	88606	30MAR61	10FEB81
1402	M	TA2	32615	17JAN63	02DEC90
1479	F	TA3	38785	22DEC68	05OCT89

Program

```
libname proclib 'SAS-library';
proc sql;
   create table proclib.bonus as
 select IdNumber, Salary format=dollar8.,
          salary*.025 as Bonus format=dollar8.
      from proclib.payroll;
title 'BONUS Information';
select *
      from proclib.bonus(obs=10);
```

Program Description

Declare the PROCLIB library. The PROCLIB library is used in these examples to store created tables.

```
libname proclib 'SAS-library';
```

Create the PROCLIB.BONUS table. The CREATE TABLE statement creates the table PROCLIB.BONUS from the result of the subsequent query.

```
proc sql;
   create table proclib.bonus as
```

Select the columns to include. The SELECT clause specifies that three columns will be in the new table: IdNumber, Salary, and Bonus. FORMAT= assigns the DOLLAR8. format to Salary. The Bonus column is built with the SQL expression *salary*.025*.

```
select IdNumber, Salary format=dollar8.,
        salary*.025 as Bonus format=dollar8.
    from proclib.payroll;
```

Specify the title.

```
title 'BONUS Information';
```

Display the first 10 rows of the PROCLIB.BONUS table. The SELECT clause selects columns from PROCLIB.BONUS. The asterisk (*) selects all columns. The FROM clause specifies PROCLIB.BONUS as the table to select from. The OBS= data set option limits the printing of the output to 10 rows.

```
select *
    from proclib.bonus(obs=10);
```

Output

Output 7.2 *Creating a Table from a Query*

BONUS Information

IdNumber	Salary	Bonus
1919	$34,376	$859
1653	$35,108	$878
1400	$29,769	$744
1350	$32,886	$822
1401	$38,822	$971
1499	$43,025	$1,076
1101	$18,723	$468
1333	$88,606	$2,215
1402	$32,615	$815
1479	$38,785	$970

Example 3: Updating Data in a PROC SQL Table

Features: ALTER TABLE statement
- DROP clause
- MODIFY clause

UPDATE statement
- SET clause

CASE expression

Table name: EMPLOYEES

This example updates data values in the EMPLOYEES table and drops a column.

Program to Create the Employee Table

```
proc sql;
   title 'Employees Table';
   select * from Employees;
```

Program Description

Display the entire EMPLOYEES table. The SELECT clause displays the table before the updates. The asterisk (*) selects all columns for display. The FROM clause specifies EMPLOYEES as the table to select from.

```
proc sql;
   title 'Employees Table';
   select * from Employees;
```

***Output 7.3** Employees Table*

Employees Table

IdNum	LName	FName	JobCode	Salary	Phone
1876	CHIN	JACK	TA1	42400	212/588-5634
1114	GREENWALD	JANICE	ME3	38000	212/588-1092
1556	PENNINGTON	MICHAEL	ME1	29860	718/383-5681
1354	PARKER	MARY	FA3	65800	914/455-2337
1130	WOOD	DEBORAH	PT2	36514	212/587-0013

Program to Update the Employee Table

```
proc sql;
update employees
      set salary=salary*
      case when jobcode like '__1' then 1.04
           else 1.025
      end;
alter table employees
      modify salary num format=dollar8.
      drop phone;
title 'Updated Employees Table';
 select * from employees;
```

Program Description

Update the values in the Salary column. The UPDATE statement updates the values in EMPLOYEES. The SET clause specifies that the data in the Salary column be multiplied by 1.04 when the job code ends with a 1 and 1.025 for all other job codes.

(The two underscores represent any character.) The CASE expression returns a value for each row that completes the SET clause.

```
proc sql;
update employees
      set salary=salary*
      case when jobcode like '__1' then 1.04
           else 1.025
      end;
```

Modify the format of the Salary column and delete the Phone column. The ALTER TABLE statement specifies EMPLOYEES as the table to alter. The MODIFY clause permanently modifies the format of the Salary column. The DROP clause permanently drops the Phone column.

```
alter table employees
      modify salary num format=dollar8.
      drop phone;
```

Specify the title.

```
title 'Updated Employees Table';
```

Display the entire updated EMPLOYEES table. The SELECT clause displays the EMPLOYEES table after the updates. The asterisk (*) selects all columns.

```
select * from employees;
```

Output

Output 7.4 *Updated Employees Table*

Updated Employees Table

IdNum	LName	FName	JobCode	Salary
1876	CHIN	JACK	TA1	$44,096
1114	GREENWALD	JANICE	ME3	$38,950
1556	PENNINGTON	MICHAEL	ME1	$31,054
1354	PARKER	MARY	FA3	$67,445
1130	WOOD	DEBORAH	PT2	$37,427

Example 4: Joining Two Tables

Features: FROM clause
- table alias

inner join

joined-table component

PROC SQL statement option
- NUMBER

WHERE clause
IN condition

Table name: PROCLIB.STAFF

Table name: PROCLIB.PAYROLL

Details

This example joins two tables in order to get more information about data that are common to both tables.

```
proc sql outobs=10;
   title 'PROCLIB.STAFF';
   title2 'First 10 Rows Only';
   select * from proclib.staff;
   title;
```

Figure 7.3 *PROCLIB.STAFF Table*

PROCLIB.STAFF
First 10 Rows Only

idnum	lname	fname	city	state	hphone
1919	ADAMS	GERALD	STAMFORD	CT	203/781-1255
1653	ALIBRANDI	MARIA	BRIDGEPORT	CT	203/675-7715
1400	ALHERTANI	ABDULLAH	NEW YORK	NY	212/586-0808
1350	ALVAREZ	MERCEDES	NEW YORK	NY	718/383-1549
1401	ALVAREZ	CARLOS	PATERSON	NJ	201/732-8787
1499	BAREFOOT	JOSEPH	PRINCETON	NJ	201/812-5665
1101	BAUCOM	WALTER	NEW YORK	NY	212/586-8060
1333	BANADYGA	JUSTIN	STAMFORD	CT	203/781-1777
1402	BLALOCK	RALPH	NEW YORK	NY	718/384-2849
1479	BALLETTI	MARIE	NEW YORK	NY	718/384-8816

```
proc sql outobs=10;
   title 'PROCLIB.PAYROLL';
   title2 'First 10 Rows Only';
   select * from proclib.payroll;
   title;
```

Figure 7.4 *PROCLIB.PAYROLL Table*

PROCLIB.PAYROLL
First 10 Rows Only

IdNumber	Gender	Jobcode	Salary	Birth	Hired
1919	M	TA2	34376	12SEP60	04JUN87
1653	F	ME2	35108	15OCT64	09AUG90
1400	M	ME1	29769	05NOV67	16OCT90
1350	F	FA3	32886	31AUG65	29JUL90
1401	M	TA3	38822	13DEC50	17NOV85
1499	M	ME3	43025	26APR54	07JUN80
1101	M	SCP	18723	06JUN62	01OCT90
1333	M	PT2	88606	30MAR61	10FEB81
1402	M	TA2	32615	17JAN63	02DEC90
1479	F	TA3	38785	22DEC68	05OCT89

Program

```
libname proclib 'SAS-library';
proc sql number;
title 'Information for Certain Employees Only';
/* Select the columns to display. The SELECT clause selects the columns to
show in the output. */
    select Lname, Fname, City, State,
           IdNumber, Salary, Jobcode
from proclib.staff, proclib.payroll
where idnumber=idnum and idnum in
              ('1919', '1400', '1350', '1333');
```

Program Description

Declare the PROCLIB library. The PROCLIB library is used in these examples to store created tables.

```
libname proclib 'SAS-library';
```

Add row numbers to PROC SQL output. NUMBER adds a column that contains the row number.

```
proc sql number;
```

Specify the title.

```
title 'Information for Certain Employees Only';
```

```
/* Select the columns to display. The SELECT clause selects the columns to
show in the output. */
    select Lname, Fname, City, State,
           IdNumber, Salary, Jobcode
```

Specify the tables from which to obtain the data. The FROM clause lists the tables to select from.

```
from proclib.staff, proclib.payroll
```

Specify the join criterion and subset the query. The WHERE clause specifies that the tables are joined on the ID number from each table. WHERE also further subsets the query with the IN condition, which returns rows for only four employees.

```
where idnumber=idnum and idnum in
              ('1919', '1400', '1350', '1333');
```

Output 7.5 *Information for Certain Employees Only*

Information for Certain Employees Only

Row	lname	fname	city	state	IdNumber	Salary	Jobcode
1	ADAMS	GERALD	STAMFORD	CT	1919	34376	TA2
2	ALHERTANI	ABDULLAH	NEW YORK	NY	1400	29769	ME1
3	ALVAREZ	MERCEDES	NEW YORK	NY	1350	32886	FA3
4	BANADYGA	JUSTIN	STAMFORD	CT	1333	88606	PT2

Example 5: Combining Two Tables

Features: DELETE statement
IS condition
RESET statement option
  DOUBLE
UNION set operator

Table name: PROCLIB.NEWPAY

Table name: PROCLIB.PAYLIST

Table name: PROCLIB.PAYLIST2

Input Tables

This example creates a new table, PROCLIB.NEWPAY, by concatenating two other tables: PROCLIB.PAYLIST and PROCLIB.PAYLIST2.

```
proc sql;
title 'PROCLIB.PAYLIST Table';
   select * from proclib.paylist;
```

Figure 7.5 *PROCLIB.PAYLIST Table*

PROCLIB.PAYLIST Table

IdNum	Gender	Jobcode	Salary	Birth	Hired
1639	F	TA1	42260	26JUN70	28JAN91
1065	M	ME3	38090	26JAN54	07JAN92
1400	M	ME1	29769	05NOV67	16OCT90
1561	M		36514	30NOV63	07OCT87
1221	F	FA3	.	22SEP63	04OCT94

```
proc sql;
title 'PROCLIB.PAYLIST2 Table';
  select * from proclib.PAYLIST2;
title;
```

Figure 7.6 *PROCLIB.PAYLIST2 Table*

PROCLIB.PAYLIST2 Table

IdNum	Gender	Jobcode	Salary	Birth	Hired
1919	M	TA2	34376	12SEP66	04JUN87
1653	F	ME2	31896	15OCT64	09AUG92
1350	F	FA3	36886	31AUG55	29JUL91
1401	M	TA3	38822	13DEC55	17NOV93
1499	M	ME1	23025	26APR74	07JUN92

Program

```
libname proclib 'SAS-library';
proc sql;
   create table proclib.newpay as
      select * from proclib.paylist
      union
      select * from proclib.paylist2;
delete
      from proclib.newpay
      where jobcode is missing or salary is missing;
reset double;
title 'Personnel Data';
select *
      from proclib.newpay;
```

Program Description

Declare the PROCLIB library. The PROCLIB library is used in these examples to store created tables.

```
libname proclib 'SAS-library';
```

Create the PROCLIB.NEWPAY table. The SELECT clauses select all the columns from the tables that are listed in the FROM clauses. The UNION set operator concatenates the query results that are produced by the two SELECT clauses.

```
proc sql;
   create table proclib.newpay as
      select * from proclib.paylist
      union
      select * from proclib.paylist2;
```

Delete rows with missing Jobcode or Salary values. The DELETE statement deletes rows from PROCLIB.NEWPAY that satisfy the WHERE expression. The IS condition specifies rows that contain missing values in the Jobcode or Salary column.

```
delete
      from proclib.newpay
      where jobcode is missing or salary is missing;
```

Reset the PROC SQL environment and double-space the output. RESET changes the procedure environment without stopping and restarting PROC SQL. The DOUBLE option double-spaces the output. (The DOUBLE option has no effect on ODS output.)

```
reset double;
```

Specify the title.

```
title 'Personnel Data';
```

Display the entire PROCLIB.NEWPAY table. The SELECT clause selects all columns from the newly created table, PROCLIB.NEWPAY.

```
select *
      from proclib.newpay;
```

Output

Output 7.6 *Personnel Data*

Personnel Data

IdNum	Gender	Jobcode	Salary	Birth	Hired
1065	M	ME3	38090	26JAN54	07JAN92
1350	F	FA3	36886	31AUG55	29JUL91
1400	M	ME1	29769	05NOV67	16OCT90
1401	M	TA3	38822	13DEC55	17NOV93
1499	M	ME1	23025	26APR74	07JUN92
1639	F	TA1	42260	26JUN70	28JAN91
1653	F	ME2	31896	15OCT64	09AUG92
1919	M	TA2	34376	12SEP66	04JUN87

Example 6: Reporting from DICTIONARY Tables

Features: DESCRIBE TABLE statement
DICTIONARY.table-name component

Table name: DICTIONARY.MEMBERS

This example uses DICTIONARY tables to show a list of the SAS files in a SAS library. If you do not know the names of the columns in the DICTIONARY table that you are querying, then use a DESCRIBE TABLE statement with the table.

Program

```
libname proclib 'SAS-library';
proc sql;
   describe table dictionary.members;
title 'SAS Files in the PROCLIB Library';
select memname, memtype
       from dictionary.members
       where libname='PROCLIB';
```

Program Description

Declare the PROCLIB library. The PROCLIB library is used in these examples to store created tables.

```
libname proclib 'SAS-library';
```

List the column names from the DICTIONARY.MEMBERS table. DESCRIBE TABLE writes the column names from DICTIONARY.MEMBERS to the SAS log.

```
proc sql;
   describe table dictionary.members;
```

Specify the title.

```
title 'SAS Files in the PROCLIB Library';
```

Display a list of files in the PROCLIB library. The SELECT clause selects the MEMNAME and MEMTYPE columns. The FROM clause specifies DICTIONARY.MEMBERS as the table to select from. The WHERE clause subsets the output to include only those rows that have a libref of *PROCLIB* in the LIBNAME column.

```
select memname, memtype
      from dictionary.members
      where libname='PROCLIB';
```

Log

```
277  options nodate pageno=1 source linesize=80 pagesize=60;
278
279  proc sql;
280     describe table dictionary.members;
NOTE: SQL table DICTIONARY.MEMBERS was created like:

create table DICTIONARY.MEMBERS
  (
   libname char(8) label='Library Name',
   memname char(32) label='Member Name',
   memtype char(8) label='Member Type',
   engine char(8) label='Engine Name',
   index char(32) label='Indexes',
   path char(1024) label='Path Name'
  );

281     title 'SAS Files in the PROCLIB Library';
282
283     select memname, memtype
284        from dictionary.members
285        where libname='PROCLIB';
```

HTML Output

Output 7.7 *SAS Files in the PROCLIB Library*

SAS Files in the PROCLIB Library

Member Name	Member Type
BONUS	DATA
NEWPAY	DATA
PAYLIST	DATA
PAYLIST2	DATA
PAYROLL	DATA
STAFF	DATA

Example 7: Performing an Outer Join

Features: joined-table component
left outer join
SELECT clause
COALESCE function
WHERE clause
CONTAINS condition

Table name: PROCLIB.PAYROLL

Table name: PROCLIB.PAYROLL2

Details

This example illustrates a left outer join of the PROCLIB.PAYROLL and PROCLIB.PAYROLL2 tables.

```
proc sql outobs=10;
   title 'PROCLIB.PAYROLL';
   title2 'First 10 Rows Only';
   select * from proclib.payroll
   order by idnumber;
   title;
```

Figure 7.7 *PROCLIB.PAYROLL*

PROCLIB.PAYROLL
First 10 Rows Only

IdNumber	Gender	Jobcode	Salary	Birth	Hired
1009	M	TA1	28880	02MAR59	26MAR92
1017	M	TA3	40858	28DEC57	16OCT81
1036	F	TA3	39392	19MAY65	23OCT84
1037	F	TA1	28558	10APR64	13SEP92
1038	F	TA1	26533	09NOV69	23NOV91
1050	M	ME2	35167	14JUL63	24AUG86
1065	M	ME2	35090	26JAN44	07JAN87
1076	M	PT1	66558	14OCT55	03OCT91
1094	M	FA1	22268	02APR70	17APR91
1100	M	BCK	25004	01DEC60	07MAY88

```
proc sql;
   title 'PROCLIB.PAYROLL2';
   select * from proclib.payroll2
    order by idnum;
   title;
```

Figure 7.8 *PROCLIB.PAYROLL2*

PROCLIB.PAYROLL2

idnum	gender	jobcode	salary	birth	hired
1036	F	TA3	42465	19MAY65	23OCT84
1065	M	ME3	38090	26JAN44	07JAN87
1076	M	PT1	69742	14OCT55	03OCT91
1106	M	PT3	94039	06NOV57	16AUG84
1129	F	ME3	36758	08DEC61	17AUG91
1221	F	FA3	29896	22SEP67	04OCT91
1350	F	FA3	36098	31AUG65	29JUL90
1369	M	TA3	36598	28DEC61	13MAR87
1447	F	FA1	22123	07AUG72	29OCT92
1561	M	TA3	36514	30NOV63	07OCT87
1639	F	TA3	42260	26JUN57	28JAN84
1998	M	SCP	23100	10SEP70	02NOV92

Program Using OUTER JOIN Based on ID Number

```
libname proclib 'SAS-library';
proc sql outobs=10;
title 'Most Current Jobcode and Salary Information';
select p.IdNumber, p.Jobcode, p.Salary,
         p2.jobcode label='New Jobcode',
         p2.salary label='New Salary' format=dollar8.
from proclib.payroll as p left join proclib.payroll2 as p2
on p.IdNumber=p2.idnum;
```

Program Description

Declare the PROCLIB library. The PROCLIB library is used in these examples to store created tables.

```
libname proclib 'SAS-library';
```

Limit the number of output rows. OUTOBS= limits the output to 10 rows.

```
proc sql outobs=10;
```

Specify the title for the first query.

```
title 'Most Current Jobcode and Salary Information';
```

Select the columns. The SELECT clause lists the columns to select. Some column names are prefixed with a table alias because they are in both tables. LABEL= and FORMAT= are column modifiers.

```
select p.IdNumber, p.Jobcode, p.Salary,
          p2.jobcode label='New Jobcode',
          p2.salary label='New Salary' format=dollar8.
```

Specify the type of join. The FROM clause lists the tables to join and assigns table aliases. The keywords LEFT JOIN specify the type of join. The order of the tables in the FROM clause is important. PROCLIB.PAYROLL is listed first and is considered the "left" table. PROCLIB.PAYROLL2 is the "right" table.

```
from proclib.payroll as p left join proclib.payroll2 as p2
```

Specify the join criterion. The ON clause specifies that the join be performed based on the values of the ID numbers from each table.

```
on p.IdNumber=p2.idnum;
```

Output for OUTER JOIN

As the output shows, all rows from the left table, PROCLIB.PAYROLL, are returned. PROC SQL assigns missing values for rows in the left table, PAYROLL, that have no matching values for IdNum in PAYROLL2.

Most Current Jobcode and Salary Information

IdNumber	Jobcode	Salary	New Jobcode	New Salary
1009	TA1	28880		.
1017	TA3	40858		.
1036	TA3	39392	TA3	$42,465
1037	TA1	28558		.
1038	TA1	26533		.
1050	ME2	35167		.
1065	ME2	35090	ME3	$38,090
1076	PT1	66558	PT1	$69,742
1094	FA1	22268		.
1100	BCK	25004		.

Program Using COALESCE and LEFT JOIN

```
proc sql outobs=10;
title 'Most Current Jobcode and Salary Information';
select p.idnumber, coalesce(p2.jobcode,p.jobcode)
        label='Current Jobcode',
coalesce(p2.salary,p.salary) label='Current Salary'
```

```
                  format=dollar8.
   from proclib.payroll p left join proclib.payroll2 p2
         on p.IdNumber=p2.idnum;
```

Program Description

```
proc sql outobs=10;
```

Specify the title for the second query.

```
title 'Most Current Jobcode and Salary Information';
```

Select the columns and coalesce the Jobcode columns. The SELECT clause lists the columns to select. COALESCE overlays the like-named columns. For each row, COALESCE returns the first nonmissing value of either P2.JOBCODE or P.JOBCODE. Because P2.JOBCODE is the first argument, if there is a nonmissing value for P2.JOBCODE, COALESCE returns that value. Thus, the output contains the most recent job code information for every employee. LABEL= assigns a column label.

```
select p.idnumber, coalesce(p2.jobcode,p.jobcode)
        label-'Current Jobcode',
```

Coalesce the Salary columns. For each row, COALESCE returns the first nonmissing value of either P2.SALARY or P.SALARY. Because P2.SALARY is the first argument, if there is a nonmissing value for P2.SALARY, then COALESCE returns that value. Thus, the output contains the most recent salary information for every employee.

```
coalesce(p2.salary,p.salary) label='Current Salary'
                  format=dollar8.
```

Specify the type of join and the join criterion. The FROM clause lists the tables to join and assigns table aliases. The keywords LEFT JOIN specify the type of join. The ON clause specifies that the join is based on the ID numbers from each table.

```
from proclib.payroll p left join proclib.payroll2 p2
      on p.IdNumber=p2.idnum;
```

Output for COALESCE and LEFT JOIN

Output 7.8 *Most Current Jobcode and Salary Information*

Most Current Jobcode and Salary Information

IdNumber	Current Jobcode	Current Salary
1009	TA1	$28,880
1017	TA3	$40,858
1036	TA3	$42,465
1037	TA1	$28,558
1038	TA1	$26,533
1050	ME2	$35,167
1065	ME3	$38,090
1076	PT1	$69,742
1094	FA1	$22,268
1100	BCK	$25,004

Program to Subset the Query

```
proc sql;

title 'Most Current Information for Ticket Agents';
   select p.IdNumber,
          coalesce(p2.jobcode,p.jobcode) label='Current Jobcode',
          coalesce(p2.salary,p.salary) label='Current Salary'
      from proclib.payroll p left join proclib.payroll2 p2
      on p.IdNumber=p2.idnum
      where p2.jobcode contains 'TA';
```

Program Description

Subset the query. The WHERE clause subsets the left join to include only those rows containing the value TA.

```
proc sql;

title 'Most Current Information for Ticket Agents';
   select p.IdNumber,
          coalesce(p2.jobcode,p.jobcode) label='Current Jobcode',
          coalesce(p2.salary,p.salary) label='Current Salary'
      from proclib.payroll p left join proclib.payroll2 p2
      on p.IdNumber=p2.idnum
      where p2.jobcode contains 'TA';
```

Output for Subset of the Query

Output 7.9 *Query Results with the Value TA*

Most Current Information for Ticket Agents

IdNumber	Current Jobcode	Current Salary
1036	TA3	42465
1369	TA3	36598
1561	TA3	36514
1639	TA3	42260

Example 8: Creating a View from a Query's Result

Features: CREATE VIEW statement
GROUP BY clause
SELECT clause
    COUNT function
HAVING clause

Other features: AVG summary function
data set option
    PW=

Table name: PROCLIB.PAYROLL

Table name: PROCLIB.JOBS

Details

This example creates the PROC SQL view PROCLIB.JOBS from the result of a query expression.

```
proc sql outobs=10;
   title 'PROCLIB.PAYROLL';
   title2 'First 10 Rows Only';
   select * from proclib.payroll
   order by idnumber;
   title;
```

Figure 7.9 PROCLIB.PAYROLL

PROCLIB.PAYROLL
First 10 Rows Only

IdNumber	Gender	Jobcode	Salary	Birth	Hired
1009	M	TA1	28880	02MAR59	26MAR92
1017	M	TA3	40858	28DEC57	16OCT81
1036	F	TA3	39392	19MAY65	23OCT84
1037	F	TA1	28558	10APR64	13SEP92
1038	F	TA1	26533	09NOV69	23NOV91
1050	M	ME2	35167	14JUL63	24AUG86
1065	M	ME2	35090	26JAN44	07JAN87
1076	M	PT1	66558	14OCT55	03OCT91
1094	M	FA1	22268	02APR70	17APR91
1100	M	BCK	25004	01DEC60	07MAY88

Program

```
libname proclib 'SAS-library';
proc sql;
   create view proclib.jobs(pw=red) as
select Jobcode,
            count(jobcode) as number label='Number',
avg(int((today()-birth)/365.25)) as avgage
               format=2. label='Average Age',
            avg(salary) as avgsal
               format=dollar8. label='Average Salary'
from payroll
group by jobcode
      having avgage ge 30;
title 'Current Summary Information for Each Job Category';
   title2 'Average Age Greater Than or Equal to 30';
select * from proclib.jobs(pw=red);
title2;
```

Program Description

Declare the PROCLIB library. The PROCLIB library is used in these examples to store created tables.

```
libname proclib 'SAS-library';
```

Create the PROCLIB.JOBS view. CREATE VIEW creates the PROC SQL view PROCLIB.JOBS. The PW= data set option assigns password protection to the data that is generated by this view.

```
proc sql;
   create view proclib.jobs(pw=red) as
```

Select the columns. The SELECT clause specifies four columns for the view: Jobcode and three columns, Number, AVGAGE, and AVGSAL, whose values are the products functions. COUNT returns the number of nonmissing values for each job code because the data is grouped by Jobcode. LABEL= assigns a label to the column.

```
select Jobcode,
            count(jobcode) as number label='Number',
```

Calculate the Avgage and Avgsal columns. The AVG summary function calculates the average age and average salary for each job code.

```
avg(int((today()-birth)/365.25)) as avgage
               format=2. label='Average Age',
            avg(salary) as avgsal
               format=dollar8. label='Average Salary'
```

Specify the table from which the data is obtained. The FROM clause specifies PAYROLL as the table to select from. PROC SQL assumes the libref of PAYROLL to be PROCLIB because PROCLIB is used in the CREATE VIEW statement.

```
from payroll
```

Organize the data into groups and specify the groups to include in the output. The GROUP BY clause groups the data by the values of Jobcode. Thus, any summary statistics are calculated for each grouping of rows by value of Jobcode. The HAVING clause subsets the grouped data and returns rows for job codes that contain an average age of greater than or equal to 30.

```
group by jobcode
      having avgage ge 30;
```

Specify the titles.

```
title 'Current Summary Information for Each Job Category';
   title2 'Average Age Greater Than or Equal to 30';
```

Display the entire PROCLIB.JOBS view. The SELECT statement selects all columns from PROCLIB.JOBS. PW=RED is necessary because the view is password protected.

```
select * from proclib.jobs(pw=red);
```

```
title2;
```

Output

Output 7.10 *View Created from the Results of a Query*

Current Summary Information for Each Job Category
Average Age Greater Than or Equal to 30

Jobcode	Number	Average Age	Average Salary
BCK	9	45	$25,794
FA1	11	42	$23,039
FA2	16	46	$27,987
FA3	7	48	$32,934
ME1	8	43	$28,500
ME2	14	49	$35,577
ME3	7	51	$42,411
NA1	5	39	$42,032
NA2	3	51	$52,383
PT1	8	47	$67,908
PT2	10	52	$87,925
PT3	2	63	$10,505
SCP	7	46	$18,309
TA1	9	45	$27,721
TA2	20	46	$33,575
TA3	12	49	$39,680

Example 9: Joining Three Tables

Features: FROM clause
joined-table component
WHERE clause

Table name: PROCLIB.STAFF2

Table name: PROCLIB.SCHEDULE2

Table name: PROCLIB.SUPERV2

Details

This example joins three tables and produces a report that contains columns from each table.

Example Code 7.1 *PROCLIB.STAFF2 Table*

```
data proclib.staff2;
input IdNum $4. @7 Lname $12. @20 Fname $8. @30 City $10.
      @42 State $2. @50 Hphone $12.;
   datalines;
1106  MARSHBURN    JASPER    STAMFORD     CT        203/781-1457
1430  DABROWSKI    SANDRA    BRIDGEPORT   CT        203/675-1647
1118  DENNIS       ROGER     NEW YORK     NY        718/383-1122
1126  KIMANI       ANNE      NEW YORK     NY        212/586-1229
1402  BLALOCK      RALPH     NEW YORK     NY        718/384-2849
1882  TUCKER       ALAN      NEW YORK     NY        718/384-0216
1479  BALLETTI     MARIE     NEW YORK     NY        718/384-8816
1420  ROUSE        JEREMY    PATERSON     NJ        201/732-9834
1403  BOWDEN       EARL      BRIDGEPORT   CT        203/675-3434
1616  FUENTAS      CARLA     NEW YORK     NY        718/384-3329
;
run;

proc sql;
   title 'PROCLIB.STAFF2';
   select * from proclib.staff2;
   title;
```

Figure 7.10 *PROCLIB.STAFF2*

PROCLIB.STAFF2

IdNum	Lname	Fname	City	State	Hphone
1106	MARSHBURN	JASPER	STAMFORD	CT	203/781-1457
1430	DABROWSKI	SANDRA	BRIDGEPORT	CT	203/675-1647
1118	DENNIS	ROGER	NEW YORK	NY	718/383-1122
1126	KIMANI	ANNE	NEW YORK	NY	212/586-1229
1402	BLALOCK	RALPH	NEW YORK	NY	718/384-2849
1882	TUCKER	ALAN	NEW YORK	NY	718/384-0216
1479	BALLETTI	MARIE	NEW YORK	NY	718/384-8816
1420	ROUSE	JEREMY	PATERSON	NJ	201/732-9834
1403	BOWDEN	EARL	BRIDGEPORT	CT	203/675-3434
1616	FUENTAS	CARLA	NEW YORK	NY	718/384-3329

Example Code 7.2 *PROCLIB.SCHEDULE2 Table*

```
data proclib.schedule2;
   input flight $3. +5 date date7. +2 dest $3. +3 idnum $4.;
   format date date7.;
   informat date date7.;
   datalines;
132     01MAR94  BOS   1118
132     01MAR94  BOS   1402
219     02MAR94  PAR   1616
219     02MAR94  PAR   1478
622     03MAR94  LON   1430
622     03MAR94  LON   1882
271     04MAR94  NYC   1430
271     04MAR94  NYC   1118
579     05MAR94  RDU   1126
579     05MAR94  RDU   1106
;
run;

proc sql;
   title 'PROCLIB.SCHEDULE2';
   select * from proclib.schedule2;
   title;
```

Figure 7.11 *PROCLIB.SCHEDULE2*

PROCLIB.SCHEDULE2

flight	date	dest	idnum
132	01MAR94	BOS	1118
132	01MAR94	BOS	1402
219	02MAR94	PAR	1616
219	02MAR94	PAR	1478
622	03MAR94	LON	1430
622	03MAR94	LON	1882
271	04MAR94	NYC	1430
271	04MAR94	NYC	1118
579	05MAR94	RDU	1126
579	05MAR94	RDU	1106

Example Code 7.3 *PROCLIB.SUPERV2 Table*

```
data proclib.superv2;
   input supid $4. +8 state $2. +5  jobcat  $2.;
   label supid='Supervisor Id' jobcat='Job Category';
```

```
   datalines;
1417       NJ     NA
1352       NY     NA
1106       CT     PT
1442       NJ     PT
1118       NY     PT
1405       NJ     SC
1564       NY     SC
1639       CT     TA
1126       NY     TA
1882       NY     ME
;
run;

proc sql;
   title 'PROCLIB.SUPERV2';
   select * from proclib.superv2
   title;
```

Figure 7.12 *PROCLIB.SUPERV2*

PROCLIB.SUPERV2

Supervisor Id	state	Job Category
1417	NJ	NA
1352	NY	NA
1106	CT	PT
1442	NJ	PT
1118	NY	PT
1405	NJ	SC
1564	NY	SC
1639	CT	TA
1126	NY	TA
1882	NY	ME

Program

```
libname proclib 'SAS-library';
proc sql;
   title 'All Flights for Each Supervisor';
   select s.IdNum, Lname, City 'Hometown', Jobcat,
          Flight, Date
from proclib.schedule2 s, proclib.staff2 t, proclib.superv2 v
where s.idnum=t.idnum and t.idnum=v.supid;
```

Program Description

Declare the PROCLIB library. The PROCLIB library is used in these examples to store created tables.

```
libname proclib 'SAS-library';
```

Select the columns. The SELECT clause specifies the columns to select. IdNum is prefixed with a table alias because it appears in two tables.

```
proc sql;
   title 'All Flights for Each Supervisor';
   select s.IdNum, Lname, City 'Hometown', Jobcat,
          Flight, Date
```

Specify the tables to include in the join. The FROM clause lists the three tables for the join and assigns an alias to each table.

```
from proclib.schedule2 s, proclib.staff2 t, proclib.superv2 v
```

Specify the join criteria. The WHERE clause specifies the columns that join the tables. The STAFF2 and SCHEDULE2 tables have an IdNum column, which has related values in both tables. The STAFF2 and SUPERV2 tables have the IdNum and SUPID columns, which have related values in both tables.

```
where s.idnum=t.idnum and t.idnum=v.supid;
```

Output

Output 7.11 *ID Values from All Three Tables Are Included*

All Flights for Each Supervisor

idnum	Lname	Hometown	Job Category	flight	date
1106	MARSHBURN	STAMFORD	PT	579	05MAR94
1118	DENNIS	NEW YORK	PT	132	01MAR94
1118	DENNIS	NEW YORK	PT	271	04MAR94
1126	KIMANI	NEW YORK	TA	579	05MAR94
1882	TUCKER	NEW YORK	ME	622	03MAR94

Example 10: Querying an In-Line View

Features: FROM clause
in-line view

Table name: PROCLIB.STAFF2

Table name: PROCLIB.SCHEDULE2

Table name: PROCLIB.SUPERV2

This example shows an alternative way to construct the query that is explained in "Example 9: Joining Three Tables" on page 268 by joining one of the tables with the results of an in-line view. The example also shows how to rename columns with an in-line view.

Program

```
libname proclib 'SAS-library';
proc sql;
   title 'All Flights for Each Supervisor';
   select three.*, v.jobcat
from (select lname, s.idnum, city, flight, date
            from proclib.schedule2 s, proclib.staff2 t
            where s.idnum=t.idnum)
as three (Surname, Emp_ID, Hometown,
                        FlightNumber, FlightDate),
proclib.superv2 v
      where three.Emp_ID=v.supid;
```

Program Description

Declare the PROCLIB library. The PROCLIB library is used in these examples to store created tables.

```
libname proclib 'SAS-library';
```

Select the columns. The SELECT clause selects all columns that are returned by the in-line view (which will have the alias Three assigned to it), plus one column from the third table (which will have the alias V assigned to it).

```
proc sql;
   title 'All Flights for Each Supervisor';
   select three.*, v.jobcat
```

Specify the in-line query. Instead of including the name of a table or view, the FROM clause includes a query that joins two of the three tables. In the in-line query, the SELECT clause lists the columns to select. IdNum is prefixed with a table alias because it appears in both tables. The FROM clause lists the two tables for the join and assigns an alias to each table. The WHERE clause specifies the columns that join the tables. The STAFF2 and SCHEDULE2 tables have an IdNum column, which has related values in both tables.

```
from (select lname, s.idnum, city, flight, date
            from proclib.schedule2 s, proclib.staff2 t
            where s.idnum=t.idnum)
```

Specify an alias for the query and names for the columns. The alias Three refers to the results of the in-line view. The names in parentheses become the names for the columns in the view.

```
as three (Surname, Emp_ID, Hometown,
                        FlightNumber, FlightDate),
```

Join the results of the in-line view with the third table. The WHERE clause specifies the columns that join the table with the in-line view. Note that the WHERE clause specifies the renamed Emp_ID column from the in-line view.

```
    proclib.superv2 v
          where three.Emp_ID=v.supid;
```

Output

Output 7.12 *Query of an In-Line View*

All Flights for Each Supervisor

Surname	Emp_ID	Hometown	FlightNumber	FlightDate	Job Category
MARSHBURN	1106	STAMFORD	579	05MAR94	PT
DENNIS	1118	NEW YORK	132	01MAR94	PT
DENNIS	1118	NEW YORK	271	04MAR94	PT
KIMANI	1126	NEW YORK	579	05MAR94	TA
TUCKER	1882	NEW YORK	622	03MAR94	ME

Example 11: Retrieving Values with the SOUNDS-LIKE Operator

Features: ORDER BY clause
SOUNDS-LIKE operator

Table name: PROCLIB.STAFF

This example returns rows based on the functionality of the SOUNDS-LIKE operator in a WHERE clause. The SOUNDS-LIKE operator is based on the SOUNDEX algorithm for identifying words that sound alike. The SOUNDEX algorithm is English-biased and is less useful for languages other than English. For more information about the "SOUNDEX Function" in *SAS Functions and CALL Routines: Reference* algorithm, see *SAS Functions and CALL Routines: Reference.*

Details

```
proc sql outobs=10;
   title 'PROCLIB.STAFF';
   title2 'First 10 Rows Only';
   select * from proclib.staff;
   title;
```

Figure 7.13 *PROCLIB.STAFF*

PROCLIB.STAFF
First 10 Rows Only

idnum	lname	fname	city	state	hphone
1919	ADAMS	GERALD	STAMFORD	CT	203/781-1255
1653	ALIBRANDI	MARIA	BRIDGEPORT	CT	203/675-7715
1400	ALHERTANI	ABDULLAH	NEW YORK	NY	212/586-0808
1350	ALVAREZ	MERCEDES	NEW YORK	NY	718/383-1549
1401	ALVAREZ	CARLOS	PATERSON	NJ	201/732-8787
1499	BAREFOOT	JOSEPH	PRINCETON	NJ	201/812-5665
1101	BAUCOM	WALTER	NEW YORK	NY	212/586-8060
1333	BANADYGA	JUSTIN	STAMFORD	CT	203/781-1777
1402	BLALOCK	RALPH	NEW YORK	NY	718/384-2849
1479	BALLETTI	MARIE	NEW YORK	NY	718/384-8816

Program to Select Names That Sound like 'Johnson'

```
libname proclib 'SAS-library';
proc sql;
   title "Employees Whose Last Name Sounds Like 'Johnson'";
   select idnum, upcase(lname), fname
      from proclib.staff
where lname=*"Johnson"
      order by 2;
```

Program Description

Declare the PROCLIB library. The PROCLIB library is used in these examples to store created tables.

```
libname proclib 'SAS-library';
```

Select the columns and the table from which the data is obtained. The SELECT clause selects all columns from the table in the FROM clause, PROCLIB.STAFF.

```
proc sql;
   title "Employees Whose Last Name Sounds Like 'Johnson'";
   select idnum, upcase(lname), fname
      from proclib.staff
```

Subset the query and sort the output. The WHERE clause uses the SOUNDS-LIKE operator to subset the table by those employees whose last name sounds like *Johnson*. The ORDER BY clause orders the output by the second column.

```
where lname=*"Johnson"
      order by 2;
```

Output for Names That Sound like 'Johnson'

Output 7.13 *Employees Whose Last Name Sounds like 'Johnson'*

Employees Whose Last Name Sounds Like 'Johnson'

idnum		fname
1411	JOHNSEN	JACK
1113	JOHNSON	LESLIE
1369	JONSON	ANTHONY

Program to Select Names That Sound like 'Sanders'

SOUNDS-LIKE is useful, but there might be instances where it does not return every row that seems to satisfy the condition. PROCLIB.STAFF has an employee with the last name SANDERS and an employee with the last name SANYERS. The algorithm does not find SANYERS, but it does find SANDERS and SANDERSON.

```
proc sql;
title "Employees Whose Last Name Sounds Like 'Sanders'";
   select *
      from proclib.staff
      where lname=*"Sanders"
      order by 2;
```

Output for Names That Sound like 'Sanders'

Output 7.14 *Employees Whose Last Name Sounds like 'Sanders'*

Employees Whose Last Name Sounds Like 'Sanders'

idnum	lname	fname	city	state	hphone
1561	SANDERS	RAYMOND	NEW YORK	NY	212/588-6615
1414	SANDERSON	NATHAN	BRIDGEPORT	CT	203/675-1715
1434	SANDERSON	EDITH	STAMFORD	CT	203/781-1333

Example 12: Joining Two Tables and Calculating a New Value

Features: GROUP BY clause

HAVING clause

SELECT clause
- ABS function
- FORMAT= column-modifier

LABEL= column-modifier
MIN summary function
** operator, exponentiation
SQRT function

Table name: STORES

Table name: HOUSES

Details

This example joins two tables in order to compare and analyze values that are unique to each table yet have a relationship with a column that is common to both tables.

```
proc sql;
   title  'STORES Table';
   title2 'Coordinates of Stores';
   select * from stores;
   title  'HOUSES Table';
   title2 'Coordinates of Houses';
   select * from houses;
title;
```

The tables contain X and Y coordinates that represent the location of the stores and houses.

Figure 7.14 *STORES and HOUSES Tables*

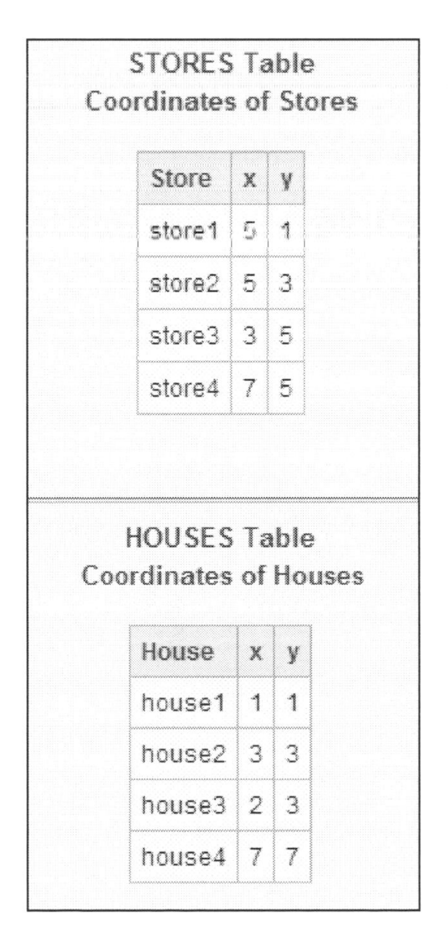

STORES Table
Coordinates of Stores

Store	x	y
store1	5	1
store2	5	3
store3	3	5
store4	7	5

HOUSES Table
Coordinates of Houses

House	x	y
house1	1	1
house2	3	3
house3	2	3
house4	7	7

Program

```
proc sql;
   title 'Each House and the Closest Store';
   select house, store label='Closest Store',
          sqrt((abs(s.x-h.x)**2)+(abs(h.y-s.y)**2)) as dist
             label='Distance' format=4.2
      from stores s, houses h
group by house
      having dist=min(dist);
```

Program Description

Specify the query. The SELECT clause specifies three columns: HOUSE, STORE, and DIST. The arithmetic expression uses the square root function (SQRT) to create the values of DIST, which contain the distance from HOUSE to STORE for each row. The double asterisk (**) represents exponentiation. LABEL= assigns a label to STORE and to DIST.

```
proc sql;
   title 'Each House and the Closest Store';
   select house, store label='Closest Store',
          sqrt((abs(s.x-h.x)**2)+(abs(h.y-s.y)**2)) as dist
             label='Distance' format=4.2
      from stores s, houses h
```

Organize the data into groups and subset the query. The minimum distance from each house to all the stores is calculated because the data are grouped by house. The HAVING clause specifies that each row be evaluated to determine whether its value of DIST is the same as the minimum distance from that house to any store.

```
group by house
      having dist=min(dist);
```

Output

Note that two stores are tied for shortest distance from house2.

Output 7.15 *New Value, Distance, Calculated from Two Tables*

Each House and the Closest Store

House	Closest Store	Distance
house1	store1	4.00
house2	store2	2.00
house2	store3	2.00
house3	store3	2.24
house4	store4	2.00

Example 13: Producing All the Possible Combinations of the Values in a Column

Features: CASE expression

joined-table component

Cross join

SELECT clause

DISTINCT keyword

Table name: PROCLIB.MARCH

Table name: FLIGHTS

Details

This example joins a table with itself to get all the possible combinations of the values in a column.

```
proc sql outobs=10;
   title 'PROCLIB.MARCH';
   title2 'First 10 Rows Only';
   select * from proclib.march;

title;
```

Figure 7.15 *PROCLIB.MARCH*

PROCLIB.MARCH
First 10 Rows Only

flight	date	depart	orig	dest	miles	boarded	capacity
114	01MAR08	7:10	LGA	LAX	2475	172	210
202	01MAR08	10:43	LGA	ORD	740	151	210
219	01MAR08	9:31	LGA	LON	3442	198	250
622	01MAR08	12:19	LGA	FRA	3857	207	250
132	01MAR08	15:35	LGA	YYZ	366	115	178
271	01MAR08	13:17	LGA	PAR	3635	138	250
302	01MAR08	20:22	LGA	WAS	229	105	180
114	02MAR08	7:10	LGA	LAX	2475	119	210
202	02MAR08	10:43	LGA	ORD	740	120	210
219	02MAR08	9:31	LGA	LON	3442	147	250

Program to Create the Flights Table

```
libname proclib 'SAS-library';
proc sql;
   create table flights as
      select distinct dest
         from proclib.march;
title 'Cities Serviced by the Airline';
select * from flights;
```

Program Description

Declare the PROCLIB library. The PROCLIB library is used in these examples to store created tables.

```
libname proclib 'SAS-library';
```

Create the FLIGHTS table. The CREATE TABLE statement creates the table FLIGHTS from the output of the query. The SELECT clause selects the unique values of Dest. DISTINCT specifies that only one row for each value of city be returned by the query and stored in the table FLIGHTS. The FROM clause specifies PROCLIB.MARCH as the table to select from.

```
proc sql;
   create table flights as
      select distinct dest
         from proclib.march;
```

Specify the title.

```
title 'Cities Serviced by the Airline';
```

Display the entire FLIGHTS table.

```
select * from flights;
```

Output for Flights Table

Output 7.16 *Cities Serviced by the Airline*

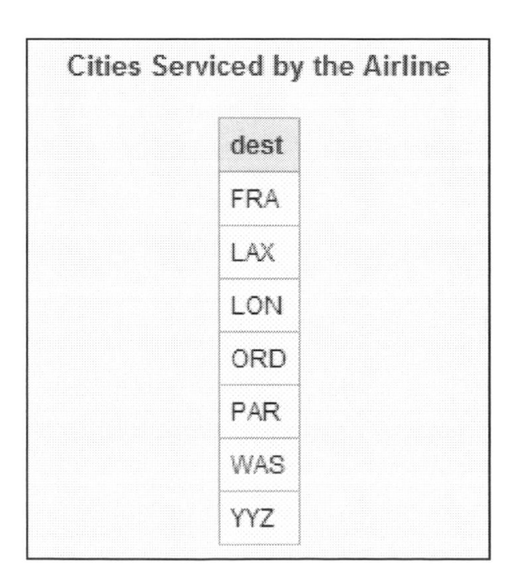
Cities Serviced by the Airline

dest
FRA
LAX
LON
ORD
PAR
WAS
YYZ

Program Using Conventional Join

```
proc sql;
title 'All Possible Connections';
select f1.Dest, case
                     when f1.dest ne ' ' then 'to and from'
                  end,
          f2.Dest
from flights as f1, flights as f2
where f1.dest < f2.dest
order by f1.dest;
```

Program Description

```
proc sql;
```

Specify the title.

```
title 'All Possible Connections';
```

Select the columns. The SELECT clause specifies three columns for the output. The prefixes on DEST are table aliases to specify which table to take the values of Dest from. The CASE expression creates a column that contains the character string to and from.

```
select f1.Dest, case
                     when f1.dest ne ' ' then 'to and from'
                  end,
          f2.Dest
```

Specify the type of join. The FROM clause joins FLIGHTS with itself and creates a table that contains every possible combination of rows (a Cartesian product). The table contains two rows for each possible route. For example, PAR <--> WAS and WAS <--> PAR.

```
from flights as f1, flights as f2
```

Specify the join criterion. The WHERE clause subsets the internal table by choosing only those rows where the name in F1.Dest sorts before the name in F2.Dest. Thus, there is only one row for each possible route.

```
where f1.dest < f2.dest
```

Sort the output. ORDER BY sorts the result by the values of F1.Dest.

```
order by f1.dest;
```

Output Using Conventional Join

Output 7.17 *All Possible Connections*

All Possible Connections

dest		dest
FRA	to and from	YYZ
FRA	to and from	WAS
FRA	to and from	LAX
FRA	to and from	ORD
FRA	to and from	PAR
FRA	to and from	LON
LAX	to and from	LON
LAX	to and from	YYZ
LAX	to and from	WAS
LAX	to and from	PAR
LAX	to and from	ORD
LON	to and from	YYZ
LON	to and from	WAS
LON	to and from	PAR
LON	to and from	ORD
ORD	to and from	YYZ
ORD	to and from	WAS
ORD	to and from	PAR
PAR	to and from	YYZ
PAR	to and from	WAS
WAS	to and from	YYZ

Program Using Cross Join

```
/*  */
proc sql;
   title 'All Possible Connections';
   select f1.Dest, case
                     when f1.dest ne ' ' then 'to and from'
                  end,
       f2.Dest
       from flights as f1 cross join flights as f2
```

```
      where f1.dest < f2.dest
      order by f1.dest;
```

Program Description

Specify a cross join. Because a cross join is functionally the same as a Cartesian product join, the cross join syntax can be substituted for the conventional join syntax.

```
/*  */
proc sql;
   title 'All Possible Connections';
   select f1.Dest, case
                    when f1.dest ne ' ' then 'to and from'
                 end,
       f2.Dest
       from flights as f1 cross join flights as f2
       where f1.dest < f2.dest
       order by f1.dest;
```

Output Using Cross Join

Output 7.18 *All Possible Connections*

All Possible Connections

dest		dest
FRA	to and from	YYZ
FRA	to and from	WAS
FRA	to and from	LAX
FRA	to and from	ORD
FRA	to and from	PAR
FRA	to and from	LON
LAX	to and from	LON
LAX	to and from	YYZ
LAX	to and from	WAS
LAX	to and from	PAR
LAX	to and from	ORD
LON	to and from	YYZ
LON	to and from	WAS
LON	to and from	PAR
LON	to and from	ORD
ORD	to and from	YYZ
ORD	to and from	WAS
ORD	to and from	PAR
PAR	to and from	YYZ
PAR	to and from	WAS
WAS	to and from	YYZ

Example 14: Matching Case Rows and Control Rows

Features: joined-table component

Table name: MATCH_11

Table name: MATCH

This example uses a table that contains data for a case-control study. Each row contains information for a case or a control. To perform statistical analysis, you need a table with one row for each case-control pair. PROC SQL joins the table with itself in order to match the cases with their appropriate controls. After the rows are matched, differencing can be performed on the appropriate columns.

The input table Appendix 3, "MATCH_11," in *Base SAS Procedures Guide* contains one row for each case and one row for each control. Pair contains a number that associates the case with its control. Low is 0 for the controls and 1 for the cases. The remaining columns contain information about the cases and controls.

```
options ls=120 nodate pageno=1;
proc sql outobs=10;
    title 'MATCH_11 Table';
    title2 'First 10 Rows Only';
    select * from match_11;
```

Figure 7.16 *MATCH_11 Table, First 10 Rows*

MATCH_11 Table
First 10 Rows Only

Pair	Low	Age	Lwt	Race	Smoke	Ptd	Ht	UI	race1	race2
1	0	14	135	1	0	0	0	0	0	0
1	1	14	101	3	1	1	0	0	0	1
2	0	15	98	2	0	0	0	0	1	0
2	1	15	115	3	0	0	0	1	0	1
3	0	16	95	3	0	0	0	0	0	1
3	1	16	130	3	0	0	0	0	0	1
4	0	17	103	3	0	0	0	0	0	1
4	1	17	130	3	1	1	0	1	0	1
5	0	17	122	1	1	0	0	0	0	0
5	1	17	110	1	1	0	0	0	0	0

Program

```
proc sql;
   create table match as
      select
         one.Low,
         one.Pair,
         (one.lwt - two.lwt) as Lwt_d,
         (one.smoke - two.smoke) as Smoke_d,
         (one.ptd - two.ptd) as Ptd_d,
         (one.ht - two.ht) as Ht_d,
         (one.ui - two.ui) as UI_d
from match_11 one, match_11 two
      where (one.pair=two.pair and one.low>two.low);
   title 'Differences for Cases and Controls';
```

```
select *
      from match(obs=5);
```

Program Description

Create the MATCH table. The SELECT clause specifies the columns for the table MATCH. SQL expressions in the SELECT clause calculate the differences for the appropriate columns and create new columns.

```
proc sql;
   create table match as
      select
         one.Low,
         one.Pair,
         (one.lwt - two.lwt) as Lwt_d,
         (one.smoke - two.smoke) as Smoke_d,
         (one.ptd - two.ptd) as Ptd_d,
         (one.ht - two.ht) as Ht_d,
         (one.ui - two.ui) as UI_d
```

Specify the type of join and the join criterion. The FROM clause lists the table MATCH_11 twice. Thus, the table is joined with itself. The WHERE clause returns only the rows for each pair that show the difference when the values for control are subtracted from the values for case.

```
from match_11 one, match_11 two
      where (one.pair=two.pair and one.low>two.low);
```

Specify the title.

```
   title 'Differences for Cases and Controls';
```

Display the first five rows of the MATCH table. The SELECT clause selects all the columns from MATCH. The OBS= data set option limits the printing of the output to five rows.

```
select *
      from match(obs=5);
```

Output

Output 7.19 *Differences for Cases and Controls*

Differences for Cases and Controls

Low	Pair	Lwt_d	Smoke_d	Ptd_d	Ht_d	UI_d
1	1	-34	1	1	0	0
1	2	17	0	0	0	1
1	3	35	0	0	0	0
1	4	27	1	1	0	1
1	5	-12	0	0	0	0

Example 15: Counting Missing Values with a SAS Macro

Features: COUNT function

Table name: SURVEY

This example uses a SAS macro to create columns. The SAS macro is not explained here. See *SAS Macro Language: Reference* for information about SAS macros.

"SURVEY" on page 388 contains data from a questionnaire about diet and exercise habits. SAS enables you to use a special notation for missing values. In the EDUC column, the **.x** notation indicates that the respondent gave an answer that is not valid, and **.n** indicates that the respondent did not answer the question. A period as a missing value indicates a data entry error.

Program

```
%macro countm(col);
    count(&col) "Valid Responses for &col",
nmiss(&col) "Missing or NOT VALID Responses for &col",
count(case
            when &col=.n  then "count me"
            end) "Coded as NO ANSWER for &col",
   count(case
            when &col=.x  then "count me"
            end) "Coded as NOT VALID answers for &col",
   count(case
            when &col=.  then "count me"
            end) "Data Entry Errors for &col"
%mend;
proc sql;
   title 'Counts for Each Type of Missing Response';
   select count(*)  "Total No. of Rows",
          %countm(educ)
      from survey;
```

Program Description

Count the nonmissing responses. The COUNTM macro uses the COUNT function to perform various counts for a column. Each COUNT function uses a CASE expression to select the rows to be counted. The first COUNT function uses only the column as an argument to return the number of nonmissing rows.

```
%macro countm(col);
    count(&col) "Valid Responses for &col",
```

Count missing or invalid responses. The NMSS function returns the number of rows for which the column has any type of missing value: .n, .x, or a period.

```
nmiss(&col) "Missing or NOT VALID Responses for &col",
```

Count the occurrences of various sources of missing or invalid responses. The last three COUNT functions use CASE expressions to count the occurrences of the three notations for missing values. The “count me” character string gives the COUNT function a nonmissing value to count.

```
count(case
            when &col=.n  then "count me"
            end) "Coded as NO ANSWER for &col",
   count(case
            when &col=.x  then "count me"
            end) "Coded as NOT VALID answers for &col",
   count(case
            when &col=.  then "count me"
            end) "Data Entry Errors for &col"
%mend;
```

Use the COUNTM macro to create the columns. The SELECT clause specifies the columns that are in the output. COUNT(*) returns the total number of rows in the table. The COUNTM macro uses the values of the EDUC column to create the columns that are defined in the macro.

```
proc sql;
   title 'Counts for Each Type of Missing Response';
   select count(*)   "Total No. of Rows",
          %countm(educ)
      from survey;
```

Output

Output 7.20 *Counts for Each Type of Missing Response*

Counts for Each Type of Missing Response

Total No. of Rows	Valid Responses for educ	Missing or NOT VALID Responses for educ	Coded as NO ANSWER for educ	Coded as NOT VALID answers for educ	Data Entry Errors for educ
8	2	6	1	3	2

Chapter 8
SQL SELECT Statement Clauses

Dictionary

SELECT Clause

Lists the columns that will appear in the output.

See: "column-definition" on page 308
"Example 1: Creating a Table and Inserting Data into It" on page 245
"Example 2: Creating a Table from a Query's Result" on page 247

Syntax

SELECT <DISTINCT> *object-item* <, ... *object-item*>

Required Arguments

alias
assigns a temporary, alternate name to the column.

DISTINCT
eliminates duplicate rows. The DISTINCT argument is identical to UNIQUE.

Note: DISTINCT works on the internal or stored value, not necessarily on the value as it is displayed. Numeric precision can cause multiple rows to be returned with values that appear to be the same.

Tips:

A row is considered a duplicate when all of its values are the same as the values of another row. The DISTINCT argument applies to all columns in the SELECT list. One row is displayed for each existing combination of values.

If available, PROC SQL uses index files when processing SELECT DISTINCT statements.

Example: "Example 13: Producing All the Possible Combinations of the Values in a Column" on page 279

object-item

is one of the following:

*

represents all columns in the tables or views that are listed in the FROM clause.

case-expression <AS *alias*>

derives a column from a CASE expression. See "CASE Expression" on page 306.

column-name<AS *alias*><column-modifier <... column-modifier>>

names a single column. See "column-name" on page 311 and "column-modifier" on page 309.

sql-expression<AS *alias*><column-modifier <... column-modifier>>

derives a column from an sql-expression. See "sql-expression" on page 338 and "column-modifier" on page 309.

table-name.*

specifies all columns in the PROC SQL table that is specified in *table-name*.

table-alias.*

specifies all columns in the PROC SQL table that has the alias that is specified in *table-alias*.

view-name.*

specifies all columns in the SAS view that is specified in *view-name*.

view-alias.*

specifies all columns in the SAS view that has the alias that is specified in *view-alias*.

UNIQUE

eliminates duplicate rows. The UNIQUE argument is identical to DISTINCT.

Note: Although the UNIQUE argument is identical to DISTINCT, it is not an ANSI standard.

Details

Asterisk (*) Notation

The asterisk (*) represents all columns of the table or tables listed in the FROM clause. When an asterisk is not prefixed with a table name, all the columns from all tables in the FROM clause are included; when it is prefixed (for example, table-name.* or table-alias.*), all the columns from that table only are included.

Note: A warning will occur if you create an output table using the SELECT * syntax when columns with the same name exist in the multiple tables that are listed on the FROM clause. You can avoid the warning by using one of the following actions:

- Individually list the desired columns in the SELECT statement at the same time as you omit the duplicate column names.

- Use the RENAME= and DROP= data set options. In this example, the ID column is renamed **tmpid**.

```
proc sql;
   create table all(drop=tmpid) as
      select * from
         one, two(rename=(id=tmpid))
            where one.id=two.tmpid;
quit;
```

If table aliases are used, place the RENAME= data set option after the table name and before the table alias. You can omit the DROP= data set option if you want to keep the renamed column in the final output table.

Column Aliases

A column alias is a temporary, alternate name for a column. Aliases are specified in the SELECT clause to name or rename columns so that the result table is clearer or easier to read. Aliases are often used to name a column that is the result of an arithmetic expression or summary function. An alias is one word only. If you need a longer column name, then use the LABEL= column-modifier, as described in "column-modifier" on page 309. The keyword AS is required with a column alias to distinguish the column alias from column names in the SELECT clause.

Column aliases are optional, and each column name in the SELECT clause can have an alias. After you assign an alias to a column, you can use the alias to refer to that column in other clauses.

If you use a column alias when creating a PROC SQL view, then the alias becomes the permanent name of the column for each execution of the view.

INTO Clause

Stores the value of one or more columns for use later in another PROC SQL query or SAS statement.

Restriction: An INTO clause cannot be used in a CREATE TABLE statement.

See: "Using the PROC SQL Automatic Macro Variables" on page 157

Syntax

INTO *macro-variable-specification*
<, ... *macro-variable-specification*>

Required Arguments

macro-variable
specifies a SAS macro variable that stores the values of the rows that are returned.

macro-variable-specification
is one of the following:

macro-variable<SEPARATED BY '*character(s)*'<NOTRIM>>
stores the values that are returned into a single macro variable.

macro-variable<TRIMMED>
stores the values that are returned into a single macro variable.

macro-variable-1 - macro-variable-n<NOTRIM>
stores the values that are returned into a range of macro variables.

Tip: When you specify a range of macro variables, the SAS Macro Facility creates only the number of macro variables that are needed. For example, if you specify `:var1-:var9999` and only 55 variables are needed, only `:var1-:var55` is created. The SQLOBS automatic variable is useful if a subsequent part of your program needs to know how many variables were actually created. In this example, SQLOBS would have the value of 55.

macro-variable-1 - <NOTRIM>
stores the values that are returned into a range of macro variables.

Tip: If you do not know how many variables you might need, you can create a macro variable range without specifying an upper bound for the range. The SQLOBS macro variable can be used if a subsequent part of your program needs to know how many variables were actually created.

NOTRIM
protects the leading and trailing blanks from being deleted from values that are stored in a range of macro variables or multiple values that are stored in a single macro variable.

SEPARATED BY '*character*'
specifies a character that separates the values of the rows.

TRIMMED
trims the leading and trailing blanks from values that are stored in a single macro variable.

Details

- Use the INTO clause only in the outer query of a SELECT statement, not in a subquery.
- When storing a value in a single macro variable, PROC SQL preserves leading or trailing blanks. The TRIMMED option can be used to trim the leading and trailing blanks from values that are stored in a single macro variable. However, if values are stored in a range of macro variables, or if the SEPARATED BY option is used to store multiple values in a single macro variable, PROC SQL trims leading or trailing blanks unless you specify the NOTRIM option.
- You can put multiple rows of the output in macro variables. You can use the PROC SQL macro variable SQLOBS to determine the number of rows that are produced by a query expression. For more information about SQLOBS, see “Using the PROC SQL Automatic Macro Variables” on page 157.

  Note: The SQLOBS macro variable is assigned a value after the SELECT statement executes.

- Values assigned by the INTO clause use the BEST12. format.

Example: INTO Clause

These examples use the “PROCLIB.HOUSES” on page 378 table:

```
title 'PROCLIB.HOUSES Table';
proc sql;
   select * from proclib.houses;
```

Output 8.1 *PROCLIB.HOUSES Table*

PROCLIB.HOUSES Table

Style	SqFeet
CONDO	900
CONDO	1000
RANCH	1200
RANCH	1400
SPLIT	1600
SPLIT	1800
TWOSTORY	2100
TWOSTORY	3000
TWOSTORY	1940
TWOSTORY	1860

With the macro-variable-specification, you can do the following:

- You can create macro variables based on the first row of the result.

```
proc sql noprint;
   select style, sqfeet
      into :style, :sqfeet
      from proclib.houses;

%put &style &sqfeet;
```

The results are written to the SAS log:

```
1   proc sql noprint;
2      select style, sqfeet
3         into :style, :sqfeet
4         from proclib.houses;
5
6   %put &style &sqfeet;
CONDO          900
```

- You can use the TRIMMED option to remove leading and trailing blanks from values that are stored in a single macro variable.

```
proc sql noprint;
   select distinct style, sqfeet
      into :s1, :s2 TRIMMED
      from proclib.houses;
%put &s1 &s2;
%put There were &sqlobs distinct values.;
```

The results are written to the SAS log:

```
1     proc sql noprint;
2       select distinct style, sqfeet
3          into :s1, :s2 TRIMMED
4          from proclib.houses;
5    %put &s1 &s2;
CONDO    900
6    %put There were &sqlobs distinct values.;
There were 1 distinct values.
```

- You can create one new macro variable per row in the result of the SELECT statement. This example shows how you can request more values for one column than for another. The hyphen is used in the INTO clause to imply a range of macro variables. You can use either of the keywords THROUGH or THRU instead of a hyphen.

  The following PROC SQL step puts the values from the first four rows of the PROCLIB.HOUSES table into macro variables:

```
proc sql noprint;
select distinct Style, SqFeet
   into :style1 - :style3, :sqfeet1 - :sqfeet4
   from proclib.houses;

%put &style1 &sqfeet1;
%put &style2 &sqfeet2;
%put &style3 &sqfeet3;
%put &sqfeet4;
```

  The %PUT statements write the results to the SAS log:

```
1   proc sql noprint;
2   select distinct style, sqfeet
3      into :style1 - :style3, :sqfeet1 - :sqfeet4
4      from proclib.houses;
5
6   %put &style1 &sqfeet1;
CONDO 900
7   %put &style2 &sqfeet2;
CONDO 1000
8   %put &style3 &sqfeet3;
RANCH 1200
9   %put &sqfeet4;
1400
```

- You can use a hyphen in the INTO clause to specify a range without an upper bound.

```
proc sql noprint;
select distinct Style, SqFeet
   into :style1 - , :sqfeet1 -
   from proclib.houses;

%put &style1 &sqfeet1;
%put &style2 &sqfeet2;
%put &style3 &sqfeet3;
%put &sqfeet4;
```

The results are written to the SAS log:

```
1 proc sql noprint;
2 select distinct Style, SqFeet
3       into :style1 - , :sqfeet1 -
4       from proclib.houses;
5
6   %put &style1 &sqfeet1;
CONDO 900
7   %put &style2 &sqfeet2;
CONDO 1000
8   %put &style3 &sqfeet3;
RANCH 1200
9   %put &sqfeet4;
1400
```

- You can concatenate the values of one column into one macro variable. This form is useful for building up a list of variables or constants. The SQLOBS macro variable is useful to reveal how many distinct variables there were in the data processed by the query.

```
proc sql noprint;
   select distinct style
      into :s1 separated by ','
      from proclib.houses;
%put &s1;
%put There were &sqlobs distinct values.;
```

The results are written to the SAS log:

```
3    proc sql noprint;
4       select distinct style
5          into :s1 separated by ','
6          from proclib.houses;
7
8   %put &s1

CONDO,RANCH,SPLIT,TWOSTORY
There were 4 distinct values.
```

- You can use leading zeros in order to create a range of macro variable names, as shown in the following example:

```
proc sql noprint;
   select SqFeet
      into :sqfeet01 - :sqfeet10
   from proclib.houses;

%put &sqfeet01 &sqfeet02 &sqfeet03 &sqfeet04 &sqfeet05;
%put &sqfeet06 &sqfeet07 &sqfeet08 &sqfeet09 &sqfeet10;
```

The results are written to the SAS log:

```
 11   proc sql noprint;
 12      select sqfeet
 13         into :sqfeet01 - :sqfeet10
 14      from proclib.houses;

15   %put &amp;sqfeet01 &amp;sqfeet02 &amp;sqfeet03 &amp;sqfeet04 &amp;sqfeet05;
900 1000 1200 1400 1600
16   %put &amp;sqfeet06 &amp;sqfeet07 &amp;sqfeet08 &amp;sqfeet09 &amp;sqfeet10;
1800 2100 3000 1940 1860
```

- You can prevent leading and trailing blanks from being trimmed from values that are stored in macro variables. By default, when storing values in a range of macro variables, or when storing multiple values in a single macro variable (with the SEPARATED BY option), PROC SQL trims the leading and trailing blanks from the values before creating the macro variables. If you do not want leading and trailing blanks to be trimmed, specify the NOTRIM option, as shown in the following example:

```
proc sql noprint;
   select style, sqfeet
      into :style1 - :style4 notrim,
           :sqfeet separated by ',' notrim
      from proclib.houses;

%put *&style1* *&sqfeet*;
%put *&style2* *&sqfeet*;
%put *&style3* *&sqfeet*;
%put *&style4* *&sqfeet*;
```

The results are written to the SAS log, as shown in the following output:

```
3   proc sql noprint;
4      select style, sqfeet
5         into :style1 - :style4 notrim,
6              :sqfeet separated by ',' notrim
7         from proclib.houses;
8
9   %put *&amp;style1* *&amp;sqfeet*;
*CONDO   * *      900,     1000,     1200,     1400,     1600,     1800,     2100,
 3000,     1940,     1860*
10  %put *&amp;style2* *&amp;sqfeet*;
*CONDO   * *      900,     1000,     1200,     1400,     1600,     1800,     2100,
 3000,     1940,     1860**
11  %put *&amp;style3* *&amp;sqfeet*;
*RANCH   * *      900,     1000,     1200,     1400,     1600,     1800,     2100,
 3000,     1940,     1860**
12  %put *&amp;style4* *&amp;sqfeet*;
*RANCH   * *      900,     1000,     1200,     1400,     1600,     1800,     2100,
 3000,     1940,     1860**</log>
</logBlock>
```

FROM Clause

Specifies source tables or views.

See: "Example 1: Creating a Table and Inserting Data into It" on page 245
"Example 4: Joining Two Tables" on page 251
"Example 9: Joining Three Tables" on page 268
"Example 10: Querying an In-Line View" on page 272

Syntax

FROM *from-list*

Required Arguments

alias
specifies a temporary, alternate name for a table, view, or in-line view that is specified in the FROM clause.

column
names the column that appears in the output. The column names that you specify are matched by position to the columns in the output.

from-list
is one of the following:

table name <<AS>*alias*>
names a single PROC SQL table. *table-name* can be a one-level name, a two-level *libref.table* name, or a physical pathname that is enclosed in single quotation marks.

view-name <<AS>*alias*>
names a single SAS view. *view-name* can be a one-level name, a two-level *libref.view* name, or a physical pathname that is enclosed in single quotation marks.

joined-table
specifies a join. See "joined-table" on page 314.

(query-expression) <<AS>*alias*> <(*column*<, ...*column*>)>
specifies an in-line view. See "query-expression" on page 330.

CONNECTION TO
specifies a DBMS table. See "CONNECTION TO" on page 311.

Note: With *table-name* and *view-name*, you can use data set options by placing them in parentheses immediately after *table-name* or *view-name*. For more information, see "Using SAS Data Set Options with PROC SQL" on page 151.

Details

Table Aliases

A table alias is a temporary, alternate name for a table that is specified in the FROM clause. Table aliases are prefixed to column names to distinguish between columns that are common to multiple tables. Column names in reflexive joins (joining a table with itself) must be prefixed with a table alias in order to distinguish which copy of the table the column comes from. Column names in other types of joins must be prefixed with table aliases or table names unless the column names are unique to those tables.

The optional keyword AS is often used to distinguish a table alias from other table names.

In-Line Views

The FROM clause can itself contain a query expression that takes an optional table alias. This type of nested query expression is called an in-line view. An in-line view is any query expression that would be valid in a CREATE VIEW statement. PROC SQL can support many levels of nesting, but it is limited to 256 tables in any one query. The 256-table limit includes underlying tables that can contribute to views that are specified in the FROM clause.

An in-line view saves you a programming step. Rather than creating a view and referring to it in another query, you can specify the view in-line in the FROM clause.

Characteristics of in-line views include the following:

- An in-line view is not assigned a permanent name, although it can take an alias.
- An in-line view can be referred to only in the query in which it is defined. It cannot be referenced in another query.
- You cannot use an ORDER BY clause in an in-line view.
- The names of columns in an in-line view can be assigned in the object-item list of that view or with a list of names enclosed in parentheses following the alias. This syntax can be useful for renaming columns. See "Example 10: Querying an In-Line View" on page 272 for an example.
- In order to visually separate an in-line view from the rest of the query, you can enclose the in-line view in any number of pairs of parentheses. Note that if you specify an alias for the in-line view, the alias specification must appear outside the outermost pair of parentheses for that in-line view.

WHERE Clause

Subsets the output based on specified conditions.

See: "Example 4: Joining Two Tables" on page 251
"Example 9: Joining Three Tables" on page 268

Syntax

WHERE sql-expression

Required Argument

sql-expression
See "sql-expression" on page 338.

Details

- When a condition is met (that is, the condition resolves to true), those rows are displayed in the result table. Otherwise, no rows are displayed.

- You cannot use summary functions that specify only one column.

  In this example, MAX is a summary function. Therefore, its context is that of a GROUP BY clause. It cannot be used to group, or summarize, data.

  ```
  where max(measure1) > 50;
  ```

  However, this WHERE clause will work.

  ```
  where max(measure1,measure2) > 50;
  ```

  In this case, MAX is a SAS function. It works with the WHERE clause because you are comparing the values of two columns within the same row. Consequently, it can be used to subset the data.

GROUP BY Clause

Specifies how to group the data for summarizing.

See: “Example 8: Creating a View from a Query's Result” on page 265

“Example 12: Joining Two Tables and Calculating a New Value” on page 276

Syntax

GROUP BY *group-by-item* <, ..., *group-by-item*>

Required Argument

group-by-item
is one of the following:

integer
is a positive integer that equates to a column's position.

column-name
is the name of a column or a column alias. See “column-name” on page 311.

sql-expression
See “sql-expression” on page 338.

Details

- You can specify more than one *group-by-item* to get more detailed reports. Both the grouping of multiple items and the BY statement of a PROC step are evaluated in similar ways. If more than one *group-by-item* is specified, then the first one determines the major grouping.
- Integers can be substituted for column names (that is, SELECT object-items) in the GROUP BY clause. For example, if the *group-by-item* is 2, then the results are grouped by the values in the second column of the SELECT clause list. Using integers can shorten your coding and enable you to group by the value of an unnamed expression in the SELECT list. Note that if you use a floating-point value (for example, 2.3), then PROC SQL ignores the decimal portion.
- The data does not have to be sorted in the order of the group-by values because PROC SQL handles sorting automatically. You can use the ORDER BY clause to specify the order in which rows are displayed in the result table.

- If you specify a GROUP BY clause in a query that does not contain a summary function, then your clause is transformed into an ORDER BY clause and a message to that effect is written to the SAS log.
- You can group the output by the values that are returned by an expression. For example, if X is a numeric variable, then the output of the following is grouped by the integer portion of values of X:

```
select x, sum(y)
from table1
group by int(x);
```

Similarly, if Y is a character variable, then the output of the following is grouped by the second character of values of Y:

```
select sum(x), y
from table1
group by substring(y from 2 for 1);
```

Note that an expression that contains only numeric literals (and functions of numeric literals) or only character literals (and functions of character literals) is ignored.

An expression in a GROUP BY clause cannot be a summary function. For example, the following GROUP BY clause is not valid:

```
group by sum(x)
```

HAVING Clause

Subsets grouped data based on specified conditions.

See: "Example 8: Creating a View from a Query's Result" on page 265 and "Example 12: Joining Two Tables and Calculating a New Value" on page 276

Syntax

HAVING sql-expression

Required Argument

sql-expression
See "sql-expression" on page 338.

Details

The HAVING clause is used with at least one summary function and an optional GROUP BY clause to summarize groups of data in a table. A HAVING clause is any valid SQL expression that is evaluated as either true or false for each group in a query. Alternatively, if the query involves remerged data, then the HAVING expression is evaluated for each row that participates in each group. The query must include one or more summary functions.

Typically, the GROUP BY clause is used with the HAVING expression and defines the group or groups to be evaluated. If you omit the GROUP BY clause, then the summary function and the HAVING clause treat the table as one group.

The following PROC SQL step uses the PROCLIB.PAYROLL table (shown in "Example 2: Creating a Table from a Query's Result" on page 247) and groups the rows by Gender to determine the oldest employee of each gender. In SAS, dates are stored as integers. The lower the birthdate as an integer, the greater the age. The expression **birth=min(birth)** is evaluated for each row in the table. When the minimum birthdate is found, the expression becomes true and the row is included in the output.

```
proc sql;
   title 'Oldest Employee of Each Gender';
   select *
      from proclib.payroll
      group by gender
      having birth=min(birth);
```

Note: This query involves remerged data because the values returned by a summary function are compared to values of a column that is not in the GROUP BY clause. See "Remerging Data" on page 350 for more information about summary functions and remerging data.

ORDER BY Clause

Specifies the order in which rows are displayed in a result table.

See: "query-expression" on page 330
"Example 11: Retrieving Values with the SOUNDS-LIKE Operator" on page 274

Syntax

ORDER BY *order-by-item* <ASC|DESC><, … *order-by-item*<ASC|DESC>>;

Required Arguments

order-by-item
is one of the following:

integer
equates to a column's position.

column-name
is the name of a column or a column alias. See "column-name" on page 311.

sql-expression
See "sql-expression" on page 338.

ASC
orders the data in ascending order. This is the default order. If neither ASC nor DESC is specified, the data is ordered in ascending order.

DESC
orders the data in descending order.

Details

- The ORDER BY clause sorts the results of a query expression according to the order specified in that query. When this clause is used, the default ordering sequence is ascending, from the lowest value to the highest. You can use the SORTSEQ= option to change the collating sequence for your output. See PROC SQL Statement on page 215.

- The order of the output rows that are returned is guaranteed only for columns that are specified in the ORDER BY clause.

  Note: The ORDER BY clause does not guarantee that the order of the rows generated is deterministic. The ANSI standard for SQL allows the SQL implementation to specify whether the ORDER BY clause is stable or unstable. If the joint combination of values that is referenced in an ORDER BY clause for a query are unique in all of the rows that are being ordered, then the order of rows that is generated by ORDER BY is always deterministic. However, if the ORDER BY clause does not reference a joint combination of unique values, then the order of rows is not deterministic if ORDER BY is unstable.

- If an ORDER BY clause is omitted, then a particular order to the output rows, such as the order in which the rows are encountered in the queried table, cannot be guaranteed—even if an index is present. Without an ORDER BY clause, the order of the output rows is determined by the internal processing of PROC SQL, the default collating sequence of SAS, and your operating environment.

- If more than one *order-by-item* is specified (separated by commas), then the first one determines the major sort order.

- Integers can be substituted for column names (that is, SELECT object-items) in the ORDER BY clause. For example, if the *order-by-item* is 2 (an integer), then the results are ordered by the values of the second column. If a query expression includes a set operator (for example, UNION), then use integers to specify the order. Doing so avoids ambiguous references to columns in the table expressions. Note that if you use a floating-point value (for example, 2.3) instead of an integer, then PROC SQL ignores the decimal portion.

- In the ORDER BY clause, you can specify any column of a table or view that is specified in the FROM clause of a query expression, regardless of whether that column has been included in the query's SELECT clause. For example, this query produces a report ordered by the descending values of the population change for each country from 1990 to 1995:

  ```
  proc sql;
     select country
        from census
        order by pop95-pop90 desc;
  ```

  ```
  NOTE: The query as specified involves
        ordering by an item that
        doesn't appear in its SELECT clause.
  ```

- You can order the output by the values that are returned by an expression. For example, if X is a numeric variable, then the output of the following is ordered by the integer portion of values of X:

  ```
  select x, y
  from table1
  order by int(x);
  ```

  Similarly, if Y is a character variable, then the output of the following is ordered by the second character of values of Y:

  ```
  select x, y
  from table1
  order by substring(y from 2 for 1);
  ```

  Note that an expression that contains only numeric literals (and functions of numeric literals) or only character literals (and functions of character literals) is ignored.

Chapter 9
SQL Procedure Components

Overview

This section describes the components that are used in SQL procedure statements. Components are the items in PROC SQL syntax that appear in roman type.

Most components are contained in clauses within the statements. For example, the basic SELECT statement includes the SELECT and FROM clauses, where each clause contains one or more components. Components can also contain other components.

For easy reference, components appear in alphabetical order, and some terms are referred to before they are defined. Use the index or the “See Also” references to refer to other statement or component descriptions that might be helpful.

Dictionary

BETWEEN Condition

Selects rows where column values are within a range of values.

Syntax

sql-expression <NOT> **BETWEEN** sql-expression
AND sql-expression

Required Argument

sql-expression
is described in "sql-expression" on page 338.

Details

- The SQL expressions must be of compatible data types. They must be either all numeric or all character types.
- Because a BETWEEN condition evaluates the boundary values as a range, it is not necessary to specify the smaller quantity first.
- You can use the NOT logical operator to exclude a range of numbers. For example, you can eliminate customer numbers between 1 and 15 (inclusive) so that you can retrieve data on more recently acquired customers.
- PROC SQL supports the same comparison operators that the DATA step supports. For example:

```
x between 1 and 3
x between 3 and 1
1<=x<=3
x>=1 and x<=3
```

BTRIM Function

Removes blanks or specified characters from the beginning, the end, or both the beginning and end of a character string.

Syntax

BTRIM (<<*btrim-specification*><'*btrim-character*' FROM>> sql-expression)

Required Arguments

btrim-specification
is one of the following:

LEADING
: removes the blanks or specified characters from the beginning of the character string.

TRAILING
: removes the blanks or specified characters from the end of the character string.

BOTH
: removes the blanks or specified characters from both the beginning and the end of the character string.

Default: BOTH

btrim-character
is a single character that is to be removed from the character string. The default character is a blank.

sql-expression
must resolve to a character string or character variable and is described in "sql-expression" on page 338.

Details

The BTRIM function operates on character strings. BTRIM removes one or more instances of a single character (the value of btrim-character) from the beginning, the end, or both the beginning and end of a string, depending whether LEADING, TRAILING, or BOTH is specified. If btrim-specification is not specified, then BOTH is used. If btrim-character is omitted, then blanks are removed.

Note: SAS adds trailing blanks to character values that are shorter than the length of the variable. Suppose you have a character variable Z, with length 10, and a value **xxabcxx**. SAS stores the value with three blanks after the last x (for a total length of 10). If you attempt to remove all the x characters with

```
btrim(both 'x' from z)
```

then the result is **abcxx** because PROC SQL sees the trailing characters as blanks, not the x character. In order to remove all the x characters, use

```
btrim(both 'x' from btrim(z))
```

The inner BTRIM function removes the trailing blanks before passing the value to the outer BTRIM function.

CALCULATED

Refers to columns already calculated in the SELECT clause.

Syntax

CALCULATED *column-alias*

Required Argument

column-alias
is the name that is assigned to the column in the SELECT clause.

Details

CALCULATED enables you to use the results of an expression in the same SELECT clause or in the WHERE clause. It is valid only when used to refer to columns that are calculated in the immediate query expression.

CASE Expression

Selects result values that satisfy specified conditions.

Examples: "Example 3: Updating Data in a PROC SQL Table" on page 249

"Example 13: Producing All the Possible Combinations of the Values in a Column" on page 279

Syntax

CASE <*case-operand*>

WHEN *when-condition* **THEN** *result-expression*

<...WHEN*when-condition*THEN*result-expression*>

<ELSE*result-expression*>

END

Required Arguments

case-operand
is a valid SQL expression that resolves to a table column whose values are compared to all the *when-conditions*. See "sql-expression" on page 338.

when-condition

- When *case-operand* is specified, *when-condition* is a shortened SQL expression that assumes *case-operand* as one of its operands and that resolves to true or false.
- When *case-operand* is not specified, *when-condition* is an SQL expression that resolves to true or false.

result-expression
is an SQL expression that resolves to a value.

Details

The CASE expression selects values if certain conditions are met. A CASE expression returns a single value that is conditionally evaluated for each row of a table (or view). Use the WHEN-THEN clauses when you want to execute a CASE expression for some but not all of the rows in the table that is being queried or created. An optional ELSE expression gives an alternative action if no THEN expression is executed.

When you omit case-operand, when-condition is evaluated as a Boolean (true or false) value. If when-condition returns a nonzero, nonmissing result, then the WHEN clause is true. If case-operand is specified, then it is compared with when-condition for equality. If case-operand equals when-condition, then the WHEN clause is true.

If the when-condition is true for the row that is being executed, then the result expression that follows THEN is executed. If when-condition is false, then PROC SQL evaluates the next when-condition until they are all evaluated. If every when-condition is false, then PROC SQL executes the ELSE expression, and its result becomes the CASE expression's result. If no ELSE expression is present and every when-condition is false, then the result of the CASE expression is a missing value.

You can use a CASE expression as an item in the SELECT clause and as either operand in an SQL expression.

Example

The following two PROC SQL steps show two equivalent CASE expressions that create a character column with the strings in the THEN clause. The CASE expression in the second PROC SQL step is a shorthand method that is useful when all the comparisons are with the same column.

```
proc sql;
   select Name, case
                when Continent = 'North America' then 'Continental U.S.'
                when Continent = 'Oceania' then 'Pacific Islands'
                else 'None'
                end as Region
      from states;

proc sql;
   select Name, case Continent
                when 'North America' then 'Continental U.S.'
                when 'Oceania' then 'Pacific Islands'
                else 'None'
                end as Region
      from states;
```

Note: When you use the shorthand method, the conditions must all be equality tests. That is, they cannot use comparison operators or other types of operators.

COALESCE Function

Returns the first nonmissing value from a list of columns.

Example: "Example 7: Performing an Outer Join" on page 259

Syntax

COALESCE (column-name <, ... column-name>)

Required Argument

column-name
is described in "column-name" on page 311.

Details

COALESCE accepts one or more column names of the same data type. The COALESCE function checks the value of each column in the order in which they are listed and returns the first nonmissing value. If only one column is listed, the COALESCE function returns the value of that column. If all the values of all arguments are missing, the COALESCE function returns a missing value.

In some SQL DBMSs, the COALESCE function is called the IFNULL function. See "PROC SQL and the ANSI Standard" on page 371 for more information.

Note: If your query contains a large number of COALESCE function calls, it might be more efficient to use a natural join instead. See "Natural Joins" on page 323.

column-definition

Defines PROC SQL's data types and dates

See: "column-modifier" on page 309

Example: "Example 1: Creating a Table and Inserting Data into It" on page 245

Syntax

column data-type <column-modifier <... column-modifier>>

Required Arguments

column
is a column name.

column-modifier
is described in "column-modifier" on page 309.

data-type
is one of the following data types:

CHARACTER|VARCHAR <(*width*)>
indicates a character column with a column width of *width*. The default column width is eight characters.

INTEGER|SMALLINT
indicates an integer column.

DECIMAL|NUMERIC|FLOAT <(*width*<, *ndec*>)>
indicates a floating-point column with a column width of *width* and *ndec* decimal places.

REAL|DOUBLE PRECISION
indicates a floating-point column.

DATE
indicates a date column.

Details

- SAS supports many but not all of the data types that SQL-based databases support.
- For all the numeric data types (INTEGER, SMALLINT, DECIMAL, NUMERIC, FLOAT, REAL, DOUBLE PRECISION, and DATE), the SQL procedure defaults to the SAS data type NUMERIC. The width and ndec arguments are ignored; PROC SQL creates all numeric columns with the maximum precision allowed by SAS. If you want to create numeric columns that use less storage space, then use the LENGTH statement in the DATA step. The various numeric data type names, along with the width and ndec arguments, are included for compatibility with other SQL software.

- For the character data types (CHARACTER and VARCHAR), the SQL procedure defaults to the SAS data type CHARACTER. The width argument is honored.
- The CHARACTER, INTEGER, and DECIMAL data types can be abbreviated to CHAR, INT, and DEC, respectively.
- A column that is declared with DATE is a SAS numeric variable with a date informat or format. You can use any of the column-modifiers to set the appropriate attributes for the column that is being defined. See *SAS Formats and Informats: Reference* for more information about dates.
- When using the VARCHAR2 data type for the Oracle database, or the VARCHAR data type for Greenplum and Aster databases, do not use trailing blanks in column values. Trailing blanks in the VARCHAR2 and VARCHAR data types are considered significant for some databases. Therefore, the results might not be correct, and the generated query is less efficient.

column-modifier

Sets column attributes.

See: "column-definition" on page 308
"SELECT Clause" on page 289

Examples: "Example 1: Creating a Table and Inserting Data into It" on page 245
"Example 2: Creating a Table from a Query's Result" on page 247

Syntax

column-modifier

Required Argument

column-modifier
is one of the following:

INFORMAT=*informatw.d*
specifies a SAS informat to be used when SAS accesses data from a table or view. You can change one permanent informat to another by using the ALTER statement. PROC SQL stores informats in its table definitions so that other SAS procedures and the DATA step can use this information when they reference tables created by PROC SQL.

See *SAS Formats and Informats: Reference* for more information about informats.

FORMAT=*formatw.d*
specifies a SAS format for determining how character and numeric values in a column are displayed by the query expression. If the FORMAT= modifier is used in the ALTER, CREATE TABLE, or CREATE VIEW statements, then it specifies the permanent format to be used when SAS displays data from that table or view. You can change one permanent format to another by using the ALTER statement.

See *SAS Formats and Informats: Reference* for more information about formats.

LABEL='*label*'

specifies a column label. If the LABEL= modifier is used in the ALTER, CREATE TABLE, or CREATE VIEW statements, then it specifies the permanent label to be used when displaying that column. You can change one permanent label to another by using the ALTER statement.

A label can begin with the following characters: a through z, A through Z, 0 through 9, an underscore (_), or a blank space. If you begin a label with any other character, such as pound sign (#), then that character is used as a split character and it splits the label onto the next line wherever it appears. For example:

```
select dropout label= '#Percentage of#Students
Who#Dropped Out' from educ(obs=5);
```

If a special character must appear as the first character in the output, then precede it with a space or a forward slash (/).

You can omit the LABEL= part of the column-modifier and still specify a label. Be sure to enclose the label in quotation marks, as in this example: `select empname "Names of Employees" from sql.employees;`

If an apostrophe must appear in the label, then type it twice so that SAS reads the apostrophe as a literal. Alternatively, you can use single and double quotation marks alternately (for example, “Date Rec'd”).

LENGTH=*length*

specifies the length of the column. This column modifier is valid only in the context of a SELECT statement.

TRANSCODE=YES|NO

for character columns, specifies whether values can be transcoded. Use TRANSCODE=NO to suppress transcoding. Note that when you create a table by using the CREATE TABLE AS statement, the transcoding attribute for a given character column in the created table is the same as it is in the source table unless you change it with the TRANSCODE= column modifier. For more information about transcoding, see *SAS National Language Support (NLS): Reference Guide.*

Default: YES

Restrictions:

The TRANSCODE=NO argument is not supported by some SAS Workspace Server clients. In SAS 9.2, if the argument is not supported, column values with TRANSCODE=NO are replaced (masked) with asterisks (*). Before SAS 9.2, column values with TRANSCODE=NO were transcoded.

Suppression of transcoding is not supported for the V6TAPE engine.

Interaction: If the TRANSCODE= attribute is set to NO for any character variable in a table, then PROC CONTENTS prints a transcode column that contains the TRANSCODE= value for each variable in the data set. If all variables in the table are set to the default TRANSCODE= value (YES), then no transcode column is printed.

Details

If you refer to a labeled column in the ORDER BY or GROUP BY clause, then you must use either the column name (not its label), the column's alias, or its ordering integer (for example, `ORDER BY 2`). See the section on SAS statements in *SAS Statements: Reference* for more information about labels.

column-name

Specifies the column to select.

See: "column-modifier" on page 309
"SELECT Clause" on page 289

Syntax

column-name

Required Argument

column-name
is one of the following:

column
is the name of a column.

table-name.column
is the name of a column in the table *table-name*.

table-alias.column
is the name of a column in the table that is referenced by *table-alias*.

view-name.column
is the name of a column in the view *view-name*.

view-alias.column
is the name of a column in the view that is referenced by *view-alias*.

Details

A column can be referred to by its name alone if it is the only column by that name in all the tables or views listed in the current query expression. If the same column name exists in more than one table or view in the query expression, then you must qualify each use of the column name by prefixing a reference to the table that contains it. Consider the following examples:

```
SALARY       /* name of the column */
EMP.SALARY   /* EMP is the table or view name */
E.SALARY     /* E is an alias for the table
                or view that contains the
                SALARY column */
```

CONNECTION TO

Retrieves and uses DBMS data in a PROC SQL query or view.

Tip: You can use CONNECTION TO in the SELECT statement's FROM clause as part of the from-list.

See: "Connecting to a DBMS by Using the SQL Procedure Pass-Through Facility" on page 166
SAS/ACCESS documentation

Syntax

CONNECTION TO *dbms-name (dbms-query)*

CONNECTION TO *alias (dbms-query)*

Required Arguments

alias
specifies an alias, if one was defined in the CONNECT statement.

dbms-name
identifies the DBMS that you are using.

dbms-query
specifies the query to send to a DBMS. The query uses the DBMS's dynamic SQL. You can use any SQL syntax that the DBMS understands, even if that syntax is not valid for PROC SQL. For example, your DBMS query can contain a semicolon.

The DBMS determines the number of tables that you can join with *dbms-query*. Each CONNECTION TO component counts as one table toward the 256-table PROC SQL limit for joins.

See *SAS/ACCESS for Relational Databases: Reference* for more information about DBMS queries.

CONTAINS Condition

Tests whether a string is part of a column's value.

Alias: ?

Restriction: The CONTAINS condition is used only with character operands.

Example: "Example 7: Performing an Outer Join" on page 259

Syntax

sql-expression <NOT> **CONTAINS** sql-expression

Required Argument

sql-expression
is described in "sql-expression" on page 338.

EXISTS Condition

Tests if a subquery returns one or more rows.

See: "Query Expressions (Subqueries)" on page 341

Syntax

<NOT> **EXISTS** (query-expression)

Required Argument

query-expression
is described in “query-expression” on page 330.

Details

The EXISTS condition is an operator whose right operand is a subquery. The result of an EXISTS condition is true if the subquery resolves to at least one row. The result of a NOT EXISTS condition is true if the subquery evaluates to zero rows. For example, the following query subsets PROCLIB.PAYROLL (which is shown in “Example 2: Creating a Table from a Query's Result” on page 247) based on the criteria in the subquery. If the value for STAFF.IDNUM is on the same row as the value **CT** in PROCLIB.STAFF (which is shown in “Example 4: Joining Two Tables” on page 251), then the matching IDNUM in PROCLIB.PAYROLL is included in the output. Thus, the query returns all the employees from PROCLIB.PAYROLL who live in **CT**.

```
proc sql;
  select *
    from proclib.payroll p
    where exists (select *
                    from proclib.staff s
                    where p.idnumber=s.idnum
                          and state='CT');
```

IN Condition

Tests set membership.

Example: “Example 4: Joining Two Tables” on page 251

Syntax

sql-expression <NOT> **IN** (query-expression | *constant* <, … *constant*>)

Required Arguments

constant
is a number or a quoted character string (or other special notation) that indicates a fixed value. Constants are also called *literals*.

query-expression
is described in “query-expression” on page 330.

sql-expression
is described in “sql-expression” on page 338.

Details

An IN condition tests if the column value that is returned by the SQL expression on the left is a member of the set (of constants or values returned by the query expression) on the right. The IN condition is true if the value of the left-hand operand is in the set of values that are defined by the right-hand operand.

IS Condition

Tests for a missing value.

Example: “Example 5: Combining Two Tables” on page 254

Syntax

sql-expression **IS** <NOT> **NULL** | **MISSING**

Required Argument

sql-expression
is described in “sql-expression” on page 338.

Details

IS NULL and IS MISSING are predicates that test for a missing value. IS NULL and IS MISSING are used in the WHERE, ON, and HAVING expressions. Each predicate resolves to true if the SQL expression's result is missing and false if it is not missing.

SAS stores a numeric missing value as a period (.) and a character missing value as a blank space. Unlike missing values in some versions of SQL, missing values in SAS always appear first in the collating sequence. Therefore, in Boolean and comparison operations, the following expressions resolve to true in a predicate:

```
 3>null
-3>null
 0>null
```

The SAS method for evaluating missing values differs from the method of the ANSI Standard for SQL. According to the Standard, these expressions are NULL. See “sql-expression” on page 338 for more information about predicates and operators. See “PROC SQL and the ANSI Standard” on page 371 for more information about the ANSI standard.

joined-table

Joins a table with itself or with other tables or views.

Restriction: Joins are limited to 256 tables.

See: “FROM Clause” on page 297
“query-expression” on page 330

Examples: “Example 4: Joining Two Tables” on page 251
“Example 7: Performing an Outer Join” on page 259
“Example 9: Joining Three Tables” on page 268
“Example 13: Producing All the Possible Combinations of the Values in a Column” on page 279
“Example 14: Matching Case Rows and Control Rows” on page 284

Syntax

table-name <<AS>*alias*>, *table-name* <<AS>*alias*>
<, ... *table-name*<<AS>*alias*>>

table-name <<AS>*alias*> <INNER> **JOIN** *table-name* <<AS>*alias*>
ON sql-expression

table-name <<AS>*alias*> **LEFT JOIN | RIGHT JOIN | FULL JOIN**
table-name <<AS>*alias*> **ON** sql-expression

table-name <<AS>*alias*> **CROSS JOIN** *table-name* <<AS>*alias*>

table-name <<AS>*alias*> **UNION JOIN** *table-name* <<AS>*alias*>

table-name <<AS>*alias*> **NATURAL**
<INNER | FULL<OUTER> | LEFT<OUTER> | RIGHT<OUTER>> **JOIN** *table-name* <<AS>*alias*>

Required Arguments

alias
specifies an alias for *table-name*. The AS keyword is optional.

sql-expression
is described in “sql-expression” on page 338.

table-name
can be one of the following:

- the name of a PROC SQL table.
- the name of a SAS view or PROC SQL view.
- a query expression. A query expression in the FROM clause is usually referred to as an inline view. See “FROM Clause” on page 297 for more information about i-line views.
- a connection to a DBMS in the form of the CONNECTION TO component. See “CONNECTION TO” on page 311 for more information.

table-name can be a one-level name, a two-level *libref.table* name, or a physical pathname that is enclosed in single quotation marks.

Note: If you include parentheses, then be sure to include them in pairs. Parentheses are not valid around comma joins (type).

Details

Types of Joins

- Inner join. See “Inner Joins” on page 316.
- Outer join. See “Outer Joins” on page 319.
- Cross join. See “Cross Joins” on page 321.
- Union join. See “Union Joins” on page 322.
- Natural join. See “Natural Joins” on page 323.

Joining Tables

When multiple tables, views, or query expressions are listed in the FROM clause, they are processed to form one table. The resulting table contains data from each contributing table. These queries are referred to as joins.

Conceptually, when two tables are specified, each row of table A is matched with all the rows of table B to produce an internal or intermediate table. The number of rows in the intermediate table (Cartesian product) is equal to the product of the number of rows in each of the source tables. The intermediate table becomes the input to the rest of the query in which some of its rows can be eliminated by the WHERE clause or summarized by a summary function.

A common type of join is an equijoin, in which the values from a column in the first table must equal the values of a column in the second table.

Table Limit

PROC SQL can process a maximum of 256 tables for a join. If you are using views in a join, then the number of tables on which the views are based count toward the 256-table limit. Each CONNECTION TO component in the pass-through facility counts as one table.

Specifying the Rows to Be Returned

The WHERE clause or ON clause contains the conditions (SQL expression) under which the rows in the Cartesian product are kept or eliminated in the result table. WHERE is used to select rows from inner joins. ON is used to select rows from inner or outer joins.

The expression is evaluated for each row from each table in the intermediate table described earlier in "Joining Tables" on page 315. The row is considered to be matching if the result of the expression is true (a nonzero, nonmissing value) for that row.

Note: You can follow the ON clause with a WHERE clause to further subset the query result. See "Example 7: Performing an Outer Join" on page 259 for an example.

Table Aliases

Table aliases are used in joins to distinguish the columns of one table from the columns in the other table or tables. A table name or alias must be prefixed to a column name when you are joining tables that have matching column names. See "FROM Clause" on page 297 for more information about table aliases.

Joining a Table with Itself

A single table can be joined with itself to produce more information. These joins are sometimes called reflexive joins. In these joins, the same table is listed twice in the FROM clause. Each instance of the table must have a table alias or you will not be able to distinguish between references to columns in either instance of the table. See "Example 13: Producing All the Possible Combinations of the Values in a Column" on page 279 and "Example 14: Matching Case Rows and Control Rows" on page 284 for examples.

Inner Joins

An inner join returns a result table for all the rows in a table that have one or more matching rows in the other tables, as specified by the SQL expression. Inner joins can be performed on up to 256 tables in the same query expression.

You can perform an inner join by using a list of table-names separated by commas or by using the INNER, JOIN, and ON keywords.

The LEFTTAB and RIGHTTAB tables are used to illustrate this type of join:

```
data lefttab;
   input Continent $ Export $ Country $;
```

```
  datalines;
NA   wheat Canada
EUR  corn  France
EUR  rice  Italy
AFR  oil   Egypt
;

data righttab;
  input Continent $ Export $ Country $;
  datalines;
NA   sugar USA
EUR  corn  Spain
EUR  beets Belgium
ASIA rice  Vietnam
;

proc sql;
   title 'Left Table - LEFTTAB';
   select * from lefttab;

   title 'Right Table - RIGHTTAB';
   select * from righttab;
```

```
              Left Table - LEFTTAB

          Continent  Export    Country
          -----------------------------
          NA         wheat     Canada
          EUR        corn      France
          EUR        rice      Italy
          AFR        oil       Egypt
```

```
              Right Table - RIGHTTAB

          Continent  Export    Country
          -----------------------------
          NA         sugar     USA
          EUR        corn      Spain
          EUR        beets     Belgium
          ASIA       rice      Vietnam
```

The following example joins the LEFTTAB and RIGHTTAB tables to get the Cartesian product of the two tables. The Cartesian product is the result of combining every row from one table with every row from another table. You get the Cartesian product when you join two tables and do not subset them with a WHERE clause or ON clause.

```
proc sql;
   title 'The Cartesian Product of';
   title2 'LEFTTAB and RIGHTTAB';
   select *
      from lefttab, righttab;
```

Output 9.1 *Cartesian Product of LEFTTAB and RIGHTTAB Tables*

The Cartesian Product of LEFTTAB and RIGHTTAB

Continent	Export	Country	Continent	Export	Country
NA	wheat	Canada	NA	sugar	USA
NA	wheat	Canada	EUR	corn	Spain
NA	wheat	Canada	EUR	beets	Belgium
NA	wheat	Canada	ASIA	rice	Vietnam
EUR	corn	France	NA	sugar	USA
EUR	corn	France	EUR	corn	Spain
EUR	corn	France	EUR	beets	Belgium
EUR	corn	France	ASIA	rice	Vietnam
EUR	rice	Italy	NA	sugar	USA
EUR	rice	Italy	EUR	corn	Spain
EUR	rice	Italy	EUR	beets	Belgium
EUR	rice	Italy	ASIA	rice	Vietnam
AFR	oil	Egypt	NA	sugar	USA
AFR	oil	Egypt	EUR	corn	Spain
AFR	oil	Egypt	EUR	beets	Belgium
AFR	oil	Egypt	ASIA	rice	Vietnam

The LEFTTAB and RIGHTTAB tables can be joined by listing the table names in the FROM clause. The following query represents an equijoin because the values of Continent from each table are matched. The column names are prefixed with the table aliases so that the correct columns can be selected.

```
proc sql;
   title 'Inner Join';
   select *
      from lefttab as l, righttab as r
      where l.continent=r.continent;
```

Output 9.2 *Inner Join*

Inner Join

Continent	Export	Country	Continent	Export	Country
NA	wheat	Canada	NA	sugar	USA
EUR	corn	France	EUR	corn	Spain
EUR	corn	France	EUR	beets	Belgium
EUR	rice	Italy	EUR	corn	Spain
EUR	rice	Italy	EUR	beets	Belgium

The following PROC SQL step is equivalent to the previous one and shows how to write an equijoin using the INNER JOIN and ON keywords.

```
proc sql;
   title 'Inner Join';
   select *
      from lefttab as l inner join
           righttab as r
      on l.continent=r.continent;
```

See Also

Examples

- "Example 4: Joining Two Tables" on page 251
- "Example 13: Producing All the Possible Combinations of the Values in a Column" on page 279
- "Example 14: Matching Case Rows and Control Rows" on page 284

Outer Joins

Outer joins are inner joins that have been augmented with rows that did not match with any row from the other table in the join. The three types of outer joins are left, right, and full.

A left outer join, specified with the keywords LEFT JOIN and ON, has all the rows from the Cartesian product of the two tables for which the SQL expression is true, plus rows from the first (LEFTTAB) table that do not match any row in the second (RIGHTTAB) table.

```
proc sql;
   title 'Left Outer Join';
   select *
      from lefttab as l left join
           righttab as r
      on l.continent=r.continent;
```

Output 9.3 *Left Outer Join*

Left Outer Join

Continent	Export	Country	Continent	Export	Country
AFR	oil	Egypt			
EUR	rice	Italy	EUR	beets	Belgium
EUR	corn	France	EUR	beets	Belgium
EUR	rice	Italy	EUR	corn	Spain
EUR	corn	France	EUR	corn	Spain
NA	wheat	Canada	NA	sugar	USA

A right outer join, specified with the keywords RIGHT JOIN and ON, has all the rows from the Cartesian product of the two tables for which the SQL expression is true, plus rows from the second (RIGHTTAB) table that do not match any row in the first (LEFTTAB) table.

```
proc sql;
   title 'Right Outer Join';
   select *
      from lefttab as l right join
           righttab as r
      on l.continent=r.continent;
```

Output 9.4 *Right Outer Join*

Right Outer Join

Continent	Export	Country	Continent	Export	Country
			ASIA	rice	Vietnam
EUR	rice	Italy	EUR	beets	Belgium
EUR	rice	Italy	EUR	corn	Spain
EUR	corn	France	EUR	beets	Belgium
EUR	corn	France	EUR	corn	Spain
NA	wheat	Canada	NA	sugar	USA

A full outer join, specified with the keywords FULL JOIN and ON, has all the rows from the Cartesian product of the two tables for which the SQL expression is true, plus rows from each table that do not match any row in the other table.

```
proc sql;
   title 'Full Outer Join';
   select *
      from lefttab as l full join
```

```
        righttab as r
    on l.continent=r.continent;
```

Output 9.5 *Full Outer Join*

Full Outer Join

Continent	Export	Country	Continent	Export	Country
AFR	oil	Egypt			
			ASIA	rice	Vietnam
EUR	rice	Italy	EUR	beets	Belgium
EUR	rice	Italy	EUR	corn	Spain
EUR	corn	France	EUR	beets	Belgium
EUR	corn	France	EUR	corn	Spain
NA	wheat	Canada	NA	sugar	USA

See Also

"Example 7: Performing an Outer Join" on page 259

Cross Joins

A cross join returns as its result table the product of the two tables.

Using the LEFTTAB and RIGHTTAB example tables, the following program demonstrates the cross join:

```
proc sql;
   title 'Cross Join';
   select *
      from lefttab as l cross join
         righttab as r;
```

Output 9.6 *Cross Join*

Cross Join

Continent	Export	Country	Continent	Export	Country
NA	wheat	Canada	NA	sugar	USA
NA	wheat	Canada	EUR	corn	Spain
NA	wheat	Canada	EUR	beets	Belgium
NA	wheat	Canada	ASIA	rice	Vietnam
EUR	corn	France	NA	sugar	USA
EUR	corn	France	EUR	corn	Spain
EUR	corn	France	EUR	beets	Belgium
EUR	corn	France	ASIA	rice	Vietnam
EUR	rice	Italy	NA	sugar	USA
EUR	rice	Italy	EUR	corn	Spain
EUR	rice	Italy	EUR	beets	Belgium
EUR	rice	Italy	ASIA	rice	Vietnam
AFR	oil	Egypt	NA	sugar	USA
AFR	oil	Egypt	EUR	corn	Spain
AFR	oil	Egypt	EUR	beets	Belgium
AFR	oil	Egypt	ASIA	rice	Vietnam

The cross join is not functionally different from a Cartesian product join. You would get the same result by submitting the following program:

```
proc sql;
   select *
      from lefttab, righttab;
```

Do not use an ON clause with a cross join. An ON clause will cause a cross join to fail. However, you can use a WHERE clause to subset the output.

Union Joins

A union join returns a union of the columns of both tables. The union join places in the results all rows with their respective column values from each input table. Columns that do not exist in one table will have null (missing) values for those rows in the result table. The following example demonstrates a union join.

```
proc sql;
   title 'Union Join';
   select *
      from lefttab union join righttab;
```

Output 9.7 *Union Join*

Union Join

Continent	Export	Country	Continent	Export	Country
			NA	sugar	USA
			EUR	corn	Spain
			EUR	beets	Belgium
			ASIA	rice	Vietnam
NA	wheat	Canada			
EUR	corn	France			
EUR	rice	Italy			
AFR	oil	Egypt			

Using a union join is similar to concatenating tables with the OUTER UNION set operator. See "query-expression" on page 330 for more information.

Do not use an ON clause with a union join. An ON clause will cause a union join to fail.

Natural Joins

A natural join selects rows from two tables that have equal values in columns that share the same name and the same type. An error results if two columns have the same name but different types. If join-specification is omitted when specifying a natural join, then INNER is implied. If no like columns are found, then a cross join is performed.

The following examples use these two tables:

```
data table1;
   input x y z;
   datalines;
1 2 3
2 1 8
6 5 4
2 5 6
;

data table2;
   input x b z;
   datalines;
1 5 3
3 5 4
2 7 8
6 0 4
;


proc sql;
   title 'table1';
   select * from table1;
```

```
      title 'table2';
      select * from table2;
   quit;
```

Output 9.8 *Tables for Natural Joins*

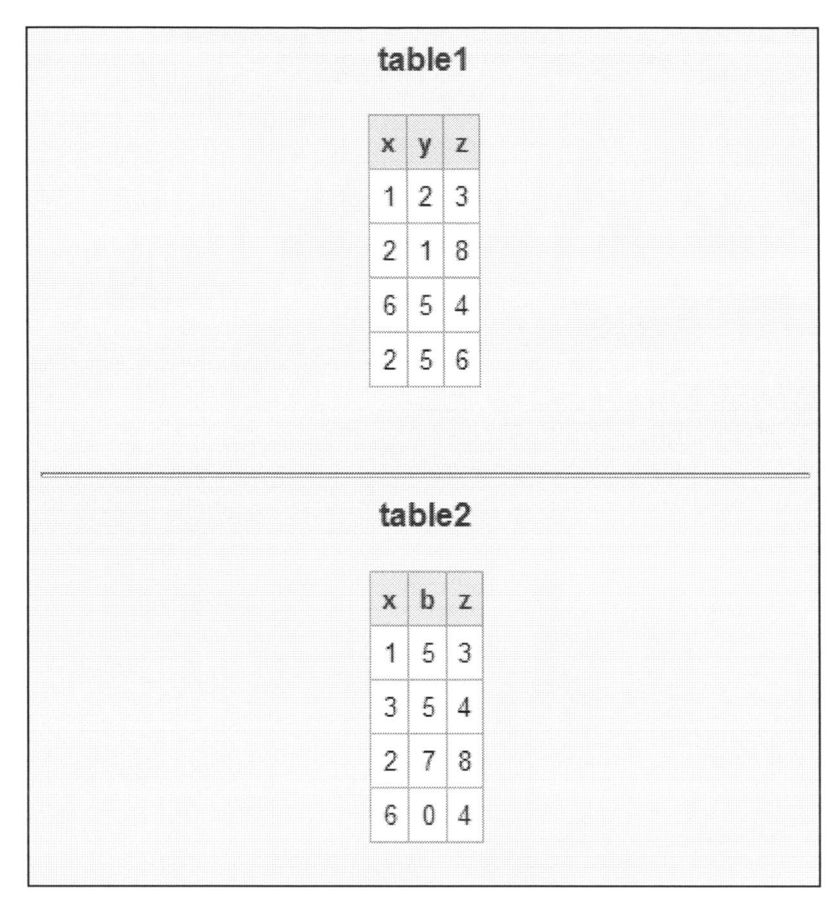

table1

x	y	z
1	2	3
2	1	8
6	5	4
2	5	6

table2

x	b	z
1	5	3
3	5	4
2	7	8
6	0	4

The following program demonstrates a natural inner join.

```
proc sql;
   title 'Natural Inner Join';
   select *
   from table1 natural join table2;
```

Output 9.9 *Natural Inner Join*

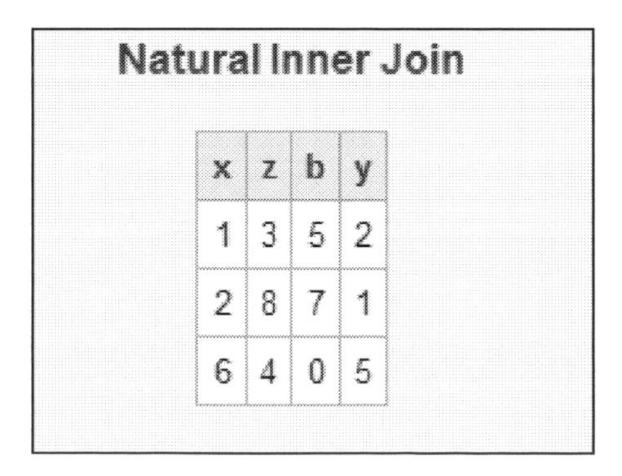

Natural Inner Join

x	z	b	y
1	3	5	2
2	8	7	1
6	4	0	5

The following program demonstrates a natural left outer join.

```
proc sql;
   title 'Natural Left Outer Join';
   select *
      from table1 natural left join table2;
```

Output 9.10 *Natural Left Outer Join*

Natural Left Outer Join

x	z	b	y
1	3	5	2
2	6	.	5
2	8	7	1
6	4	0	5

Do not use an ON clause with a natural join. An ON clause will cause a natural join to fail. When using a natural join, an ON clause is implied, matching all like columns.

Joining More than Two Tables

Inner joins are usually performed on two or three tables, but they can be performed on up to 256 tables in PROC SQL. You can combine several joins of the same or different types as shown in the following code lines:

```
a natural join b natural join c

a natural join b cross join c
```

You can also use parentheses to group joins together and control what joins happen in what order as shown in the following examples:

```
(a, b) left join c on a.X=c.Y

a left join (b full join c on b.Z-c.Z) on a.Y-b.Y
```

Note: Commutative behavior varies depending on the type of join that is performed.

A join on three tables is described here to explain how and why the relationships work among the tables.

In a three-way join, the SQL expression consists of two conditions: one condition relates the first table to the second table; and the other condition relates the second table to the third table. It is possible to break this example into stages. You could perform a two-way join to create a temporary table and then you could join the temporary table with the third one. However, PROC SQL can do it all in one step as shown in the next example. The final table would be the same in both cases.

The example shows the joining of three tables: COMM, PRICE, and AMOUNT. To calculate the total revenue from exports for each country, you need to multiply the amount exported (AMOUNT table) by the price of each unit (PRICE table), and you must know the commodity that each country exports (COMM table).

```
data comm;
   input Continent $ Export $ Country $ ;
   datalines;
```

```
NA   wheat Canada
EUR  corn  France
EUR  rice  Italy
AFR  oil   Egypt
;

data price;
   input Export $ Price;
   datalines;
rice 3.56
corn 3.45
oil  18
wheat 2.98
;

data amount;
   input Country $ Quantity;
   datalines;
Canada 16000
France  2400
Italy    500
Egypt  10000
;


proc sql;
   title 'COMM Table';
   select * from comm;
   title 'PRICE Table';
   select * from price;
   title 'AMOUNT Table';
   select * from amount;
```

Output 9.11 *Source for Joining More than Two Tables*

COMM Table

Continent	Export	Country
NA	wheat	Canada
EUR	corn	France
EUR	rice	Italy
AFR	oil	Egypt

PRICE Table

Export	Price
rice	3.56
corn	3.45
oil	18
wheat	2.98

AMOUNT Table

Country	Quantity
Canada	16000
France	2400
Italy	500
Egypt	10000

```
proc sql;
title 'Total Export Revenue';
select c.Country, p.Export, p.Price,
      a.Quantity, a.quantity*p.price
      as Total
   from comm as c JOIN price as p
       on (c.export=p.export)
        JOIN amount as a
      on (c.country=a.country);
quit;
```

Output 9.12 *Three-Way Join*

Total Export Revenue

Country	Export	Price	Quantity	Total
Canada	wheat	2.98	16000	47680
France	corn	3.45	2400	8280
Italy	rice	3.56	500	1780
Egypt	oil	18	10000	180000

See Also

“Example 9: Joining Three Tables” on page 268

Comparison of Joins and Subqueries

You can often use a subquery or a join to get the same result. However, it is often more efficient to use a join if the outer query and the subquery do not return duplicate rows. For example, the following queries produce the same result. The second query is more efficient:

```
proc sql;
   select IDNumber, Birth
      from proclib.payroll
      where IDNumber in (select idnum
                         from proclib.staff
                         where lname like 'B%');

proc sql;
   select  p.IDNumber, p.Birth
     from proclib.payroll p, proclib.staff s
     where p.idnumber=s.idnum
           and s.lname like 'B%';
```

Note: PROCLIB.PAYROLL is shown in “Example 2: Creating a Table from a Query's Result” on page 247.

LIKE Condition

Tests for a matching pattern.

Syntax

sql-expression <NOT> **LIKE** sql-expression <ESCAPE *character-expression*>

Required Arguments

sql-expression
is described in “sql-expression” on page 338.

character-expression
is an SQL expression that evaluates to a single character. The operands of *character-expression* must be character or string literals.

Note: If you use an ESCAPE clause, then the pattern-matching specification must be a quoted string or quoted concatenated string; it cannot contain column names.

Details

The LIKE condition selects rows by comparing character strings with a pattern-matching specification. It resolves to true and displays the matched strings if the left operand matches the pattern specified by the right operand. The ESCAPE clause is used to search for literal instances of the percent (%) and underscore (_) characters, which are usually used for pattern matching.

Patterns for Searching

Patterns consist of three classes of characters:

underscore (_)
matches any single character.

percent sign (%)
matches any sequence of zero or more characters.

any other character
matches that character.

These patterns can appear before, after, or on both sides of characters that you want to match. The LIKE condition is case-sensitive.

The following list uses these values: `Smith`, `Smooth`, `Smothers`, `Smart`, and `Smuggle`.

`'Sm%'`
matches `Smith`, `Smooth`, `Smothers`, `Smart`, `Smuggle`.

`'%th'`
matches `Smith`, `Smooth`.

`'S__gg%'`
matches `Smuggle`.

`'S_o'`
matches a three-letter word, so it has no matches here.

`'S_o%'`
matches `Smooth`, `Smothers`.

`'S%th'`
matches `Smith`, `Smooth`.

`'Z'`
matches the single, uppercase character `Z` only, so it has no matches here.

Searching for Literal % and _

Because the % and _ characters have special meaning in the context of the LIKE condition, you must use the ESCAPE clause to search for these character literals in the input character string.

These examples use the values `app`, `a_%`, `a__`, `bbaa1`, and `ba_1`.

- The condition `like 'a_%'` matches `app`, `a_%`, and `a__`, because the underscore (_) in the search pattern matches any single character (including the underscore), and the percent (%) in the search pattern matches zero or more characters, including '%' and '_'.
- The condition `like 'a_^%' escape '^'` matches only `a_%`, because the escape character (^) specifies that the pattern search for a literal '%'.
- The condition `like 'a_%' escape '_'` matches none of the values, because the escape character (_) specifies that the pattern search for an 'a' followed by a literal '%', which does not apply to any of these values.

Searching for Mixed-Case Strings

To search for mixed-case strings, use the UPCASE function to make all the names uppercase before entering the LIKE condition:

```
upcase(name) like 'SM%';
```

Note: When you are using the % character, be aware of the effect of trailing blanks. You might have to use the TRIM function to remove trailing blanks in order to match values.

LOWER Function

Converts the case of a character string to lowercase.

See: “UPPER Function” on page 356

Syntax

LOWER (sql-expression)

Required Argument

sql-expression
must resolve to a character string and is described in “sql-expression” on page 338.

Details

The LOWER function operates on character strings. LOWER changes the case of its argument to all lowercase.

Note: The LOWER function is provided for compatibility with the ANSI SQL standard. You can also use the SAS function LOWCASE.

query-expression

Retrieves data from tables.

See: “table-expression” on page 355
“Query Expressions (Subqueries)” on page 341
“In-Line Views” on page 298

Syntax

table-expression <*set-operator* table-expression> <...*set-operator* table-expression>

Required Arguments

table-expression
is described in "table-expression" on page 355.

set-operator
is one of the following:

INTERSECT <CORRESPONDING> <ALL>

OUTER UNION <CORRESPONDING>

UNION <CORRESPONDING> <ALL>

EXCEPT <CORRESPONDING> <ALL>

Details

Query Expressions and Table Expressions

A query expression is one or more table expressions. Multiple table expressions are linked by set operators. The following figure illustrates the relationship between table expressions and query expressions.

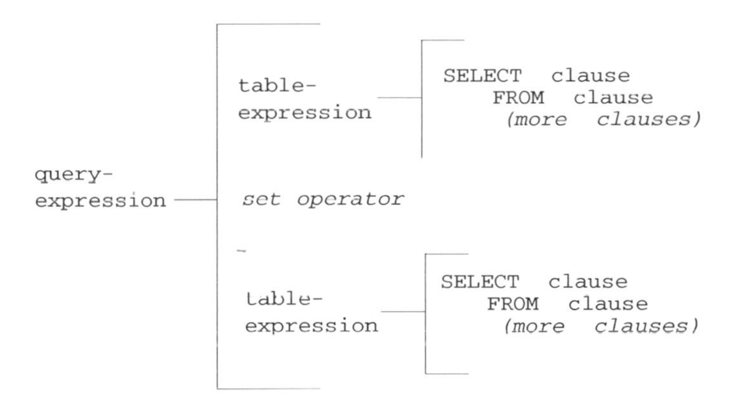

Set Operators

PROC SQL provides these set operators:

OUTER UNION
concatenates the query results.

UNION
produces all unique rows from both queries.

EXCEPT
produces rows that are part of the first query only.

INTERSECT
produces rows that are common to both query results.

A query expression with set operators is evaluated as follows.

- Each table expression is evaluated to produce an (internal) intermediate result table.
- Each intermediate result table then becomes an operand linked with a set operator to form an expression. For example, A UNION B.

- If the query expression involves more than two table expressions, then the result from the first two becomes an operand for the next set operator and operand, such as (A UNION B) EXCEPT C, ((A UNION B) EXCEPT C) INTERSECT D, and so on.
- Evaluating a query expression produces a single output table.

Set operators follow this order of precedence unless they are overridden by parentheses in the expressions: INTERSECT is evaluated first. OUTER UNION, UNION, and EXCEPT have the same level of precedence.

PROC SQL performs set operations even if the tables or views that are referred to in the table expressions do not have the same number of columns. The reason for this behavior is that the ANSI Standard for SQL requires that tables or views that are involved in a set operation have the same number of columns and that the columns have matching data types. If a set operation is performed on a table or view that has fewer columns than the one or ones with which it is being linked, then PROC SQL extends the table or view with fewer columns by creating columns with missing values of the appropriate data type. This temporary alteration enables the set operation to be performed correctly.

CORRESPONDING (CORR) Keyword

The CORRESPONDING keyword is used only when a set operator is specified. CORR causes PROC SQL to match the columns in table expressions by name and not by ordinal position. Columns that do not match by name are excluded from the result table, except for the OUTER UNION operator. See "OUTER UNION" on page 332.

For example, when performing a set operation on two table expressions, PROC SQL matches the first specified column-name (listed in the SELECT clause) from one table expression with the first specified column-name from the other. If CORR is omitted, then PROC SQL matches the columns by ordinal position.

ALL Keyword

The set operators automatically eliminate duplicate rows from their output tables. The optional ALL keyword preserves the duplicate rows, reduces the execution by one step, and thereby improves the query expression's performance. You use it when you want to display all the rows resulting from the table expressions, rather than just the unique rows. The ALL keyword is used only when a set operator is also specified.

OUTER UNION

Performing an OUTER UNION is very similar to performing the SAS DATA step with a SET statement. The OUTER UNION concatenates the intermediate results from the table expressions. Thus, the result table for the query expression contains all the rows produced by the first table expression followed by all the rows produced by the second table expression. Columns with the same name are in separate columns in the result table.

For example, the following query expression concatenates the ME1 and ME2 tables but does not overlay like-named columns. Output 9.14 on page 334 shows the result.

```
data me1;
   input IDnum $ Jobcode $ Salary Bonus;
   datalines;
1400  ME1        29769  587
1403  ME1        28072  342
1120  ME1        28619  986
1120  ME1        28619  986
;
```

```
data me2;
   input IDnum $ Jobcode $ Salary;
   datalines;
1653   ME2         35108
1782   ME2         35345
1244   ME2         36925
;

proc sql ;
   title 'ME1';
   select * from me1;
   title 'ME2';
   select * from me2;
```

Output 9.13 *ME1 and ME2 Tables*

ME1

IDnum	Jobcode	Salary	Bonus
1400	ME1	29769	587
1403	ME1	28072	342
1120	ME1	28619	986
1120	ME1	28619	986

ME2

IDnum	Jobcode	Salary
1653	ME2	35108
1782	ME2	35345
1244	ME2	36925

```
proc sql;
   title 'ME1 and ME2: OUTER UNION';
   select *
      from me1
   outer union
   select *
      from me2;
```

Output 9.14 *Outer Union of ME1 and ME2 Tables*

ME1 and ME2: OUTER UNION

IDnum	Jobcode	Salary	Bonus	IDnum	Jobcode	Salary
1400	ME1	29769	587			.
1403	ME1	28072	342			.
1120	ME1	28619	986			.
1120	ME1	28619	986			.
		.	.	1653	ME2	35108
		.	.	1782	ME2	35345
		.	.	1244	ME2	36925

Concatenating tables with the OUTER UNION set operator is similar to performing a union join. See “Union Joins” on page 322 for more information.

To overlay columns with the same name, use the CORRESPONDING keyword.

```
proc sql;
   title 'ME1 and ME2: OUTER UNION CORRESPONDING';
   select *
      from me1
   outer union corr
   select *
      from me2;
```

Output 9.15 *Outer Union Corresponding*

ME1 and ME2: OUTER UNION CORRESPONDING

IDnum	Jobcode	Salary	Bonus
1400	ME1	29769	587
1403	ME1	28072	342
1120	ME1	28619	986
1120	ME1	28619	986
1653	ME2	35108	.
1782	ME2	35345	.
1244	ME2	36925	.

In the resulting concatenated table, notice the following:

- OUTER UNION CORRESPONDING retains all nonmatching columns.

- For columns with the same name, if a value is missing from the result of the first table expression, then the value in that column from the second table expression is inserted.
- The ALL keyword is not used with OUTER UNION because this operator's default action is to include all rows in a result table. Thus, both rows from the table ME1 where IDnum is `1120` appear in the output.

UNION

The UNION operator produces a table that contains all the unique rows that result from both table expressions. That is, the output table contains rows produced by the first table expression, the second table expression, or both.

Columns are appended by position in the tables, regardless of the column names. However, the data type of the corresponding columns must match or the union will not occur. PROC SQL issues a warning message and stops executing.

The names of the columns in the output table are the names of the columns from the first table expression unless a column (such as an expression) has no name in the first table expression. In such a case, the name of that column in the output table is the name of the respective column in the second table expression.

In the following example, PROC SQL combines the two tables:

```
proc sql;
   title 'ME1 and ME2: UNION';
   select *
      from me1
   union
   select *
      from me2;
```

Output 9.16 *Union of ME1 and ME2 Tables*

ME1 and ME2: UNION

IDnum	Jobcode	Salary	Bonus
1120	ME1	28619	986
1244	ME2	36925	.
1400	ME1	29769	587
1403	ME1	28072	342
1653	ME2	35108	.
1782	ME2	35345	.

In the following example, ALL includes the duplicate row from ME1. In addition, ALL changes the sorting by specifying that PROC SQL make one pass only. Thus, the values from ME2 are simply appended to the values from ME1.

```
proc sql;
   title 'ME1 and ME2: UNION ALL';
   select *
      from me1
```

```
union all
select *
   from me2;
```

Output 9.17 *Union All*

ME1 and ME2: UNION ALL

IDnum	Jobcode	Salary	Bonus
1400	ME1	29769	587
1403	ME1	28072	342
1120	ME1	28619	986
1120	ME1	28619	986
1653	ME2	35108	.
1782	ME2	35345	.
1244	ME2	36925	.

See “Example 5: Combining Two Tables” on page 254 for another example.

EXCEPT

The EXCEPT operator produces (from the first table expression) an output table that has unique rows that are not in the second table expression. If the intermediate result from the first table expression has at least one occurrence of a row that is not in the intermediate result of the second table expression, then that row (from the first table expression) is included in the result table.

In the following example, the IN_USA table contains flights to cities within and outside the USA. The OUT_USA table contains flights only to cities outside the USA.

```
data in_usa;
   input Flight $ Dest $;
   datalines;
145 ORD
156 WAS
188 LAX
193 FRA
207 LON
;
data OUT_USA;
   input Flight $ Dest $;
   datalines;
193 FRA
207 LON
311 SJA
;

proc sql;
   title 'IN_USA';
   select * from in_usa;
```

```
    title 'OUT_USA';
    select * from out_usa;
```

Output 9.18 *Source Tables for Except Examples*

IN_USA

Flight	Dest
145	ORD
156	WAS
188	LAX
193	FRA
207	LON

OUT_USA

Flight	Dest
193	FRA
207	LON
311	SJA

This example returns only the rows from IN_USA that are not also in OUT_USA:

```
proc sql;
   title 'Flights from IN_USA Only';
   select * from in_usa
   except
   select * from out_usa;
```

Output 9.19 *Flights from IN_USA Only*

Flights from IN_USA Only

Flight	Dest
145	ORD
156	WAS
188	LAX

INTERSECT

The INTERSECT operator produces an output table that has rows that are common to both tables. For example, using the IN_USA and OUT_USA tables shown above, the following example returns rows that are in both tables:

```
proc sql;
    title 'Flights from Both IN_USA and OUT_USA';
    select * from in_usa
    intersect
    select * from out_usa;
```

Output 9.20 *Flights from Both IN_USA and OUT_USA*

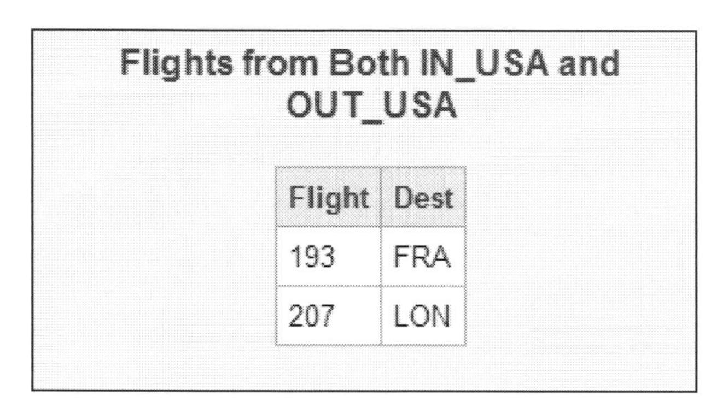

Flights from Both IN_USA and OUT_USA

Flight	Dest
193	FRA
207	LON

sql-expression

Produces a value from a sequence of operands and operators.

Syntax

operand operator operand

Required Arguments

operand

is one of the following:

- a *constant*, which is a number or a quoted character string (or other special notation) that indicates a fixed value. Constants are also called *literals*. Constants are described in *SAS Functions and CALL Routines: Reference*.
- a column-name, which is described in “column-name” on page 311.
- a CASE expression, which is described in “CASE Expression” on page 306.
- any supported SAS function. PROC SQL supports many of the functions available to the SAS DATA step. Some of the functions that are not supported are the variable information functions, functions that work with arrays of data, and functions that operate on rows other than the current row. Other SQL databases support their own sets of functions. Functions are described in the *SAS Functions and CALL Routines: Reference*.
- any functions, except those with array elements, that are created with PROC FCMP.

- the ANSI SQL functions COALESCE, BTRIM, LOWER, UPPER, and SUBSTRING.
- a summary-function, which is described in "summary-function" on page 347.
- a query expression, which is described in "query-expression" on page 330.
- the USER literal, which references the user ID of the person who submitted the program. The user ID that is returned is operating environment-dependent, but PROC SQL uses the same value that the &SYSJOBID macro variable has on the operating environment.

operator
is described in "Operators and the Order of Evaluation" on page 339.

Note: SAS functions, including summary functions, can stand alone as SQL expressions. For example

```
select min(x) from table;

   select scan(y,4) from table;
```

Details

SAS Functions

PROC SQL supports many of the functions available to the SAS DATA step. Some of the functions that are not supported are the variable information functions and functions that work with arrays of data. Other SQL databases support their own sets of functions. For example, the SCAN function is used in the following query:

```
select style, scan(street,1) format=$15.
   from houses;
```

PROC SQL also supports any user-written functions, except those functions with array elements, that are created using Chapter 18, "FCMP Procedure" in *Base SAS Procedures Guide*.

See the *SAS Functions and CALL Routines: Reference* for complete documentation of SAS functions. Summary functions are also SAS functions. For more information, see "summary-function" on page 347.

USER Literal

USER can be specified in a view definition. For example, you can create a view that restricts access to the views in the user's department. Note that the USER literal value is stored in uppercase, so it is advisable to use the UPCASE function when comparing to this value:

```
create view myemp as
   select * from dept12.employees
      where upcase(manager)=user;
```

This view produces a different set of employee information for each manager who references it.

Operators and the Order of Evaluation

The order in which operations are evaluated is the same as in the DATA step with this one exception: NOT is grouped with the logical operators AND and OR in PROC SQL; in the DATA step, NOT is grouped with the unary plus and minus signs.

Unlike missing values in some versions of SQL, missing values in SAS always appear first in the collating sequence. Therefore, in Boolean and comparison operations, the following expressions resolve to true in a predicate:

```
 3>null
-3>null
 0>null
```

You can use parentheses to group values or to nest mathematical expressions. Parentheses make expressions easier to read and can also be used to change the order of evaluation of the operators. Evaluating expressions with parentheses begins at the deepest level of parentheses and moves outward. For example, SAS evaluates A+B*C as A+(B*C), although you can add parentheses to make it evaluate as (A+B)*C for a different result.

Higher priority operations are performed first: that is, group 0 operators are evaluated before group 5 operators. The following table shows the operators and their order of evaluation, including their priority groups.

Table 9.1 *Operators and Order of Evaluation*

<table>
<tr><th>Group</th><th>Operator</th><th>Description</th></tr>
<tr><td>0</td><td>( )</td><td>forces the expression enclosed to be evaluated first</td></tr>
<tr><td>1</td><td>case-expression</td><td>selects result values that satisfy specified conditions</td></tr>
<tr><td>2</td><td>**</td><td>raises to a power</td></tr>
<tr><td></td><td>unary +, unary -</td><td>indicates a positive or negative number</td></tr>
<tr><td>3</td><td>*</td><td>multiplies</td></tr>
<tr><td></td><td>/</td><td>divides</td></tr>
<tr><td>4</td><td>+</td><td>adds</td></tr>
<tr><td></td><td>–</td><td>subtracts</td></tr>
<tr><td>5</td><td>||</td><td>concatenates</td></tr>
<tr><td>6</td><td><NOT> BETWEEN condition</td><td>See “BETWEEN Condition” on page 304.</td></tr>
<tr><td></td><td><NOT> CONTAINS condition</td><td>see “CONTAINS Condition” on page 312.</td></tr>
<tr><td></td><td><NOT> EXISTS condition</td><td>See “EXISTS Condition” on page 312.</td></tr>
<tr><td></td><td><NOT> IN condition</td><td>See “IN Condition” on page 313.</td></tr>
<tr><td></td><td>IS <NOT> condition</td><td>See “IS Condition” on page 314.</td></tr>
<tr><td></td><td><NOT> LIKE condition</td><td>See “LIKE Condition” on page 328.</td></tr>
<tr><td>7</td><td>=, eq</td><td>equals</td></tr>
<tr><td></td><td>¬=, ^=, < >, ne</td><td>does not equal</td></tr>
</table>

Group	Operator	Description
	>, gt	is greater than
	<, lt	is less than
	>=, ge	is greater than or equal to
	<=, le	is less than or equal to
	=*	sounds like (use with character operands only). See "Example 11: Retrieving Values with the SOUNDS-LIKE Operator" on page 274.
	eqt	equal to truncated strings (use with character operands only). See "Truncated String Comparison Operators" on page 341.
	gtt	greater than truncated strings
	ltt	less than truncated strings
	get	greater than or equal to truncated strings
	let	less than or equal to truncated strings
	net	not equal to truncated strings
8	¬, ^, NOT	indicates logical NOT
9	&, AND	indicates logical AND
10	|, OR	indicates logical OR

Symbols for operators might vary, depending on your operating environment. For more information, see "SAS Operators in Expressions" in Chapter 6 of *SAS Language Reference: Concepts*.

Truncated String Comparison Operators

PROC SQL supports truncated string comparison operators. (See Group 7 in Table 9.1 on page 340.) In a truncated string comparison, the comparison is performed after making the strings the same length by truncating the longer string to be the same length as the shorter string. For example, the expression `'TWOSTORY' eqt 'TWO'` is true because the string 'TWOSTORY' is reduced to 'TWO' before the comparison is performed. Note that the truncation is performed internally; neither operand is permanently changed.

Note: Unlike the DATA step, PROC SQL does not support the colon operators (such as =:, >:, and <=:) for truncated string comparisons. Use the alphabetic operators (such as EQT, GTT, and LET).

Query Expressions (Subqueries)

A query expression is called a subquery when it is used in a WHERE or HAVING clause. A subquery is a query expression that is nested as part of another query expression. A subquery selects one or more rows from a table based on values in another table.

Depending on the clause that contains it, a subquery can return a single value or multiple values. If more than one subquery is used in a query expression, then the innermost query is evaluated first, then the next innermost query, and so on, moving outward.

PROC SQL allows a subquery (contained in parentheses) at any point in an expression where a simple column value or constant can be used. In this case, a subquery must return a single value, that is, one row with only one column.

The following is an example of a subquery that returns one value. This PROC SQL step subsets the PROCLIB.PAYROLL table based on information in the PROCLIB.STAFF table. (PROCLIB.PAYROLL is shown in “Example 2: Creating a Table from a Query's Result” on page 247, and PROCLIB.STAFF is shown in “Example 4: Joining Two Tables” on page 251.) PROCLIB.PAYROLL contains employee identification numbers (IdNumber) and their salaries (Salary) but does not contain their names. If you want to return only the row from PROCLIB.PAYROLL for one employee, then you can use a subquery that queries the PROCLIB.STAFF table, which contains the employees' identification numbers and their names (Lname and Fname).

```
proc sql;
    title 'Information for Earl Bowden';
    select *
       from proclib.payroll
       where idnumber=
             (select idnum
             from proclib.staff
             where upcase(lname)='BOWDEN');
```

Output 9.21 *Query Output – One Value*

Information for Earl Bowden

IdNumber	Gender	Jobcode	Salary	Birth	Hired
1403	M	ME1	28072	28JAN69	21DEC91

Subqueries can return multiple values. The following example uses the tables PROCLIB.DELAY and PROCLIB.MARCH. These tables contain information about the same flights and have the Flight column in common. The following subquery returns all the values for Flight in PROCLIB.DELAY for international flights. The values from the subquery complete the WHERE clause in the outer query. Thus, when the outer query is executed, only the international flights from PROCLIB.MARCH are in the output.

```
proc sql outobs=5;
   title 'International Flights from';
   title2 'PROCLIB.MARCH';
   select Flight, Date, Dest, Boarded
      from proclib.march
      where flight in
            (select flight
             from proclib.delay
             where destype='International');
```

***Output* 9.22** *Query Output – Multiple Values*

International Flights from PROCLIB.MARCH

flight	date	dest	boarded
219	01MAR08	LON	198
622	01MAR08	FRA	207
132	01MAR08	YYZ	115
271	01MAR08	PAR	138
219	02MAR08	LON	147

Sometimes it is helpful to compare a value with a set of values returned by a subquery. The keywords ANY or ALL can be specified before a subquery when the subquery is the right-hand operand of a comparison. If ALL is specified, then the comparison is true only if it is true for all values that are returned by the subquery. If a subquery returns no rows, then the result of an ALL comparison is true for each row of the outer query.

If ANY is specified, then the comparison is true if it is true for any one of the values that are returned by the subquery. If a subquery returns no rows, then the result of an ANY comparison is false for each row of the outer query.

The following example selects all of the employees in PROCLIB.PAYROLL who earn more than the highest paid **ME3**:

```
proc sql;
title "Employees who Earn More than";
title2 "All ME's";
   select *
     from proclib.payroll
     where salary > all (select salary
                           from proclib.payroll
                           where jobcode='ME3');
```

Output 9.23 *Query Output Using ALL Comparison*

Employees who Earn More than All ME's

IdNumber	Gender	Jobcode	Salary	Birth	Hired
1333	M	PT2	88606	30MAR61	10FEB81
1739	M	PT1	66517	25DEC64	27JAN91
1428	F	PT1	68767	04APR60	16NOV91
1404	M	PT2	91376	24FEB53	01JAN80
1935	F	NA2	51081	28MAR54	16OCT81
1905	M	PT1	65111	16APR72	29MAY92
1407	M	PT1	68096	23MAR69	18MAR90
1410	M	PT2	84685	03MAY67	07NOV86
1439	F	PT1	70736	06MAR64	10SEP90
1545	M	PT1	66130	12AUG59	29MAY90
1106	M	PT2	89632	06NOV57	16AUG84
1442	F	PT2	84536	05SEP66	12APR88
1417	M	NA2	52270	27JUN64	07MAR89
1478	M	PT2	84203	09AUG59	24OCT90
1556	M	PT1	71349	22JUN64	11DEC91
1352	M	NA2	53798	02DEC60	16OCT86
1890	M	PT2	91908	20JUL51	25NOV79
1107	M	PT2	89977	09JUN54	10FEB79
1830	F	PT2	84471	27MAY57	29JAN83
1928	M	PT2	89858	16SEP54	13JUL90
1076	M	PT1	66558	14OCT55	03OCT91

Note: See the first item in “Subqueries and Efficiency” on page 345 for a note about efficiency when using ALL.

In order to visually separate a subquery from the rest of the query, you can enclose the subquery in any number of pairs of parentheses.

Correlated Subqueries

In a correlated subquery, the WHERE expression in a subquery refers to values in a table in the outer query. The correlated subquery is evaluated for each row in the outer query. With correlated subqueries, PROC SQL executes the subquery and the outer query together.

The following example uses the PROCLIB.DELAY and PROCLIB.MARCH tables. A DATA step ("PROCLIB.DELAY" on page 377) creates PROCLIB.DELAY. PROCLIB.MARCH is shown in "Example 13: Producing All the Possible Combinations of the Values in a Column" on page 279. PROCLIB.DELAY has the Flight, Date, Orig, and Dest columns in common with PROCLIB.MARCH:

```
proc sql outobs=5;
   title 'International Flights';
   select *
      from proclib.march
      where 'International' in
           (select destype
            from proclib.delay
            where march.Flight=delay.Flight);
```

The subquery resolves by substituting every value for MARCH.Flight into the subquery's WHERE clause, one row at a time. For example, when MARCH.Flight=**219**, the subquery resolves as follows:

1. PROC SQL retrieves all the rows from DELAY where Flight=**219** and passes their DESTYPE values to the WHERE clause.
2. PROC SQL uses the DESTYPE values to complete the WHERE clause:

   ```
   where 'International' in
      ('International','International', ...)
   ```

3. The WHERE clause checks to determine whether **International** is in the list. Because it is, all rows from MARCH that have a value of **219** for Flight become part of the output.

The following output contains the rows from MARCH for international flights only.

Output 9.24 *Correlated Subquery Output*

International Flights

flight	date	depart	orig	dest	miles	boarded	capacity
219	01MAR08	9:31	LGA	LON	3442	198	250
622	01MAR08	12:19	LGA	FRA	3857	207	250
132	01MAR08	15:35	LGA	YYZ	366	115	178
271	01MAR08	13:17	LGA	PAR	3635	138	250
219	02MAR08	9:31	LGA	LON	3442	147	250

Subqueries and Efficiency

- Use the MAX function in a subquery instead of the ALL keyword before the subquery. For example, the following queries produce the same result, but the second query is more efficient:

  ```
  proc sql;
     select * from proclib.payroll
     where salary> all(select salary
                       from proclib.payroll
  ```

```
                              where jobcode='ME3');

proc sql;
   select * from proclib.payroll
   where salary> (select max(salary)
                    from proclib.payroll
                    where jobcode='ME3');
```

- With subqueries, use IN instead of EXISTS when possible. For example, the following queries produce the same result, but the second query is usually more efficient:

```
proc sql;
   select *
      from proclib.payroll p
      where exists (select *
                     from staff s
                     where p.idnum=s.idnum
                        and state='CT');

proc sql;
   select *
      from proclib.payroll
      where idnum in (select idnum
                        from staff
                        where state='CT');
```

SUBSTRING Function

Returns a part of a character expression.

Syntax

SUBSTRING (sql-expression FROM *start* <FOR *length*>)

Required Arguments

sql-expression
: must be a character string and is described in "sql-expression" on page 338.

start
: is a number (not a variable or column name) that specifies the position, counting from the left end of the character string, at which to begin extracting the substring.

length
: is a number (not a variable or column name) that specifies the length of the substring that is to be extracted.

Details

The SUBSTRING function operates on character strings. SUBSTRING returns a specified part of the input character string, beginning at the position that is specified by start. If length is omitted, then the SUBSTRING function returns all characters from start to the end of the input character string. The values of start and length must be numbers (not variables) and can be positive, negative, or zero.

If start is greater than the length of the input character string, then the SUBSTRING function returns a zero-length string.

If start is less than 1, then the SUBSTRING function begins extraction at the beginning of the input character string.

If length is specified, then the sum of start and length cannot be less than start or an error is returned. If the sum of start and length is greater than the length of the input character string, then the SUBSTRING function returns all characters from start to the end of the input character string. If the sum of start and length is less than 1, then the SUBSTRING function returns a zero-length string.

Note: The SUBSTRING function is provided for compatibility with the ANSI SQL standard. You can also use the SAS function SUBSTR.

summary-function

Performs statistical summary calculations.

Restriction: A summary function cannot appear in an ON clause or a WHERE clause.

See: "GROUP BY Clause" on page 299
"HAVING Clause" on page 300
"SELECT Clause" on page 289
"table-expression" on page 355

Examples: "Example 8: Creating a View from a Query's Result" on page 265
"Example 12: Joining Two Tables and Calculating a New Value" on page 276
"Example 15: Counting Missing Values with a SAS Macro" on page 287

Syntax

summary-function (<DISTINCT | ALL> sql-expression)

Required Arguments

summary-function
is one of the following:

AVG|MEAN
arithmetic mean or average of values

COUNT|FREQ|N
number of nonmissing values

CSS
corrected sum of squares

CV
coefficient of variation (percent)

MAX
largest value

MIN
smallest value

NMISS
: number of missing values

PRT
: is the two-tailed *p*-value for Student's *t* statistic, T with $n - 1$ degrees of freedom.

RANGE
: range of values

STD
: standard deviation

STDERR
: standard error of the mean

SUM
: sum of values

SUMWGT
: sum of the WEIGHT variable values[1]

T
: Student's *t* value for testing the hypothesis that the population mean is zero

USS
: uncorrected sum of squares

VAR
: variance

For a description and the formulas used for these statistics, see “SAS Elementary Statistics Procedures” in Chapter 1 of *Base SAS Procedures Guide.*

DISTINCT
specifies that only the unique values of an SQL expression be used in the calculation.

ALL
specifies that all values of an SQL expression be used in the calculation. If neither DISTINCT nor ALL is specified, then ALL is used.

sql-expression
is described in “sql-expression” on page 338.

Details

Summarizing Data

Summary functions produce a statistical summary of the entire table or view that is listed in the FROM clause or for each group that is specified in a GROUP BY clause. If GROUP BY is omitted, then all the rows in the table or view are considered to be a single group. These functions reduce all the values in each row or column in a table to one summarizing or aggregate value. For this reason, these functions are often called aggregate functions. For example, the sum (one value) of a column results from the addition of all the values in the column.

Counting Rows

The COUNT function counts rows. COUNT(*) returns the total number of rows in a group or in a table. If you use a column name as an argument to COUNT, then the result is the total number of rows in a group or in a table that have a nonmissing value for that

1 Currently, there is no way to designate a WEIGHT variable for a table in PROC SQL. Thus, each row (or observation) has a weight of 1.

column. If you want to count the unique values in a column, then specify COUNT(DISTINCT *column*).

If the SELECT clause of a table expression contains one or more summary functions and that table expression resolves to no rows, then the summary function results are missing values. The following are exceptions that return zeros:

- COUNT(*)
- COUNT(<DISTINCT> sql-expression)
- NMISS(<DISTINCT> sql-expression)

See "Example 8: Creating a View from a Query's Result" on page 265 and "Example 15: Counting Missing Values with a SAS Macro" on page 287 for examples.

Calculating Statistics Based on the Number of Arguments

The number of arguments that is specified in a summary function affects how the calculation is performed. If you specify a single argument, then the values in the column are calculated. If you specify multiple arguments, then the arguments or columns that are listed are calculated for each row.

Note: When more than one argument is used within an SQL aggregate function, the function is no longer considered to be an SQL aggregate or summary function. If there is a like-named Base SAS function, then PROC SQL executes the Base SAS function, and the results that are returned are based on the values for the current row. If no like-named Base SAS function exists, then an error will occur. For example, if you use multiple arguments for the AVG function, an error will occur because there is no AVG function for Base SAS.

For example, consider calculations on the following table.

```
data summary;
   input X Y Z;
   datalines;
1 3 4
2 4 5
8 9 4
4 5 4
;

proc sql;
   title 'Summary Table';
   select * from summary;
```

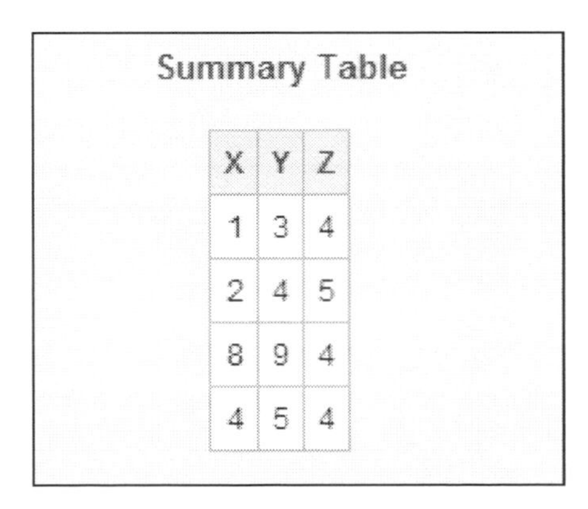

Summary Table

X	Y	Z
1	3	4
2	4	5
8	9	4
4	5	4

If you use one argument in the function, then the calculation is performed on that column only. If you use more than one argument, then the calculation is performed on each row of the specified columns. In the following PROC SQL step, the MIN and MAX functions return the minimum and maximum of the columns that they are used with. The SUM function returns the sum of each row of the columns specified as arguments:

```
proc sql;
    select min(x) as Colmin_x,
           min(y) as Colmin_y,
           max(z) as Colmax_z,
           sum(x,y,z) as Rowsum
       from summary;
```

Output 9.25 *Summary Functions*

Summary Table

Colmin_x	Colmin_y	Colmax_z	Rowsum
1	3	5	8
1	3	5	11
1	3	5	21
1	3	5	13

Remerging Data

When you use a summary function in a SELECT clause or a HAVING clause, you might see the following message in the SAS log:

```
NOTE: The query requires remerging summary
      statistics back with the original
      data.
```

The process of remerging involves two passes through the data. On the first pass, PROC SQL

- calculates and returns the value of summary functions. It then uses the result to calculate the arithmetic expressions in which the summary function participates.
- groups data according to the GROUP BY clause.

On the second pass, PROC SQL retrieves any additional columns and rows that it needs to show in the output.

Note: To specify that PROC SQL not process queries that use remerging of data, use either the PROC SQL NOREMERGE option or the NOSQLREMERGE system option. If remerging is attempted when the NOMERGE option or the NOSQLREMERGE system option is set, an error is written to the SAS log. For more information, see the REMERGE|NOREMERGE on page 222 and the “SQLREMERGE System Option” on page 368.

The following examples use the PROCLIB.PAYROLL table (shown in “Example 2: Creating a Table from a Query's Result” on page 247) to show when remerging of data is and is not necessary.

The first query requires remerging. The first pass through the data groups the data by Jobcode and resolves the AVG function for each group. However, PROC SQL must make a second pass in order to retrieve the values of IdNumber and Salary.

```
proc sql outobs=10;
   title 'Salary Information';
   title2 '(First 10 Rows Only)';
   select  IdNumber, Jobcode, Salary,
           avg(salary) as AvgSalary
      from proclib.payroll
      group by jobcode;
```

Output 9.26 *Salary Information That Required Remerging*

Salary Information
(First 10 Rows Only)

IdNumber	Jobcode	Salary	AvgSalary
1704	BCK	25465	25794.22
1677	BCK	26007	25794.22
1383	BCK	25823	25794.22
1845	BCK	25996	25794.22
1100	BCK	25004	25794.22
1663	BCK	26452	25794.22
1673	BCK	25477	25794.22
1389	BCK	25028	25794.22
1834	BCK	26896	25794.22
1132	FA1	22413	23039.36

You can change the previous query to return only the average salary for each jobcode. The following query does not require remerging because the first pass of the data does the summarizing and the grouping. A second pass is not necessary.

```
proc sql outobs=10;
   title 'Average Salary for Each Jobcode';
   select Jobcode, avg(salary) as AvgSalary
   from proclib.payroll
   group by jobcode;
```

Output 9.27 *Salary Information That Did Not Require Remerging*

Average Salary for Each Jobcode

Jobcode	AvgSalary
BCK	25794.22
FA1	23039.36
FA2	27986.88
FA3	32933.86
ME1	28500.25
ME2	35576.86
ME3	42410.71
NA1	42032.2
NA2	52383
PT1	67908

When you use the HAVING clause, PROC SQL might have to remerge data to resolve the HAVING expression.

First, consider a query that uses HAVING but that does not require remerging. The query groups the data by values of Jobcode, and the result contains one row for each value of Jobcode and summary information for people in each Jobcode. On the first pass, the summary functions provide values for the **Number**, **Average Age**, and **Average Salary** columns. The first pass provides everything that PROC SQL needs to resolve the HAVING clause, so no remerging is necessary.

```
proc sql outobs=10;
title 'Summary Information for Each Jobcode';
title2 '(First 10 Rows Only)';
   select Jobcode,
          count(jobcode) as number
              label='Number',
          avg(int((today()-birth)/365.25))
              as avgage format=2.
              label='Average Age',
          avg(salary) as avgsal format=dollar8.
              label='Average Salary'
      from proclib.payroll
      group by jobcode
      having avgage ge 30;
```

Output 9.28 *Jobcode Information That Did Not Require Remerging*

Summary Information for Each Jobcode
(First 10 Rows Only)

Jobcode	Number	Average Age	Average Salary
BCK	9	46	$25,794
FA1	11	42	$23,039
FA2	16	46	$27,987
FA3	7	48	$32,934
ME1	8	43	$28,500
ME2	14	49	$35,577
ME3	7	51	$42,411
NA1	5	39	$42,032
NA2	3	51	$52,383
PT1	8	47	$67,908

In the following query, PROC SQL remerges the data because the HAVING clause uses the SALARY column in the comparison and SALARY is not in the GROUP BY clause.

```
proc sql outobs=10;
title 'Employees who Earn More than the';
title2 'Average for Their Jobcode';
title3 '(First 10 Rows Only)';
   select Jobcode, Salary,
          avg(salary) as AvgSalary
     from proclib.payroll
     group by jobcode
     having salary > AvgSalary;
```

Output 9.29 *Jobcode Information That Did Require Remerging*

Employees who Earn More than the Average for Their Jobcode (First 10 Rows Only)

Jobcode	Salary	AvgSalary
BCK	26007	25794.22
BCK	25823	25794.22
BCK	25996	25794.22
BCK	26452	25794.22
BCK	26896	25794.22
FA1	23177	23039.36
FA1	23738	23039.36
FA1	23979	23039.36
FA1	23916	23039.36
FA1	23644	23039.36

Keep in mind that PROC SQL remerges data when

- the values returned by a summary function are used in a calculation. For example, the following query returns the values of X and the percentage of the total for each row. On the first pass, PROC SQL computes the sum of X, and on the second pass PROC SQL computes the percentage of the total for each value of X:

```
data summary;
   input x;
   datalines;
32
86
49
49
;

proc sql;
   title 'Percentage of the Total';
   select X, (100*x/sum(X)) as Pct_Total
      from summary;
```

```
                Percentage of the Total

                        x  Pct_Total
                  -------------------
                       32   14.81481
                       86   39.81481
                       49   22.68519
                       49   22.68519
```

- the values returned by a summary function are compared to values of a column that is not specified in the GROUP BY clause. For example, the following query uses the PROCLIB.PAYROLL table. PROC SQL remerges data because the column Salary is not specified in the GROUP BY clause:

```
proc sql;
   select  jobcode,  salary,
           avg(salary) as avsal
      from proclib.payroll
      group by jobcode
      having salary > avsal;
```

- a column from the input table is specified in the SELECT clause and is not specified in the GROUP BY clause. This rule does not refer to columns used as arguments to summary functions in the SELECT clause.

  For example, in the following query, the presence of IdNumber in the SELECT clause causes PROC SQL to remerge the data because IdNumber is not involved in grouping or summarizing during the first pass. In order for PROC SQL to retrieve the values for IdNumber, it must make a second pass through the data.

```
proc sql;
   select IdNumber, jobcode,
          avg(salary) as avsal
      from proclib.payroll
      group by jobcode;
```

table-expression

Defines part or all of a query expression.

See: "query-expression" on page 330
SELECT Statement on page 243

Syntax

SELECT <DISTINCT> *object-item*<, ... *object-item*>

<INTO :*macro-variable-specification*<, ... :*macro-variable-specification*>>

FROM *from-list*

<WHERE sql-expression>

<GROUP BY*group-by-item*<, ... *group-by-item*>>

<HAVING sql-expression>

Details

A table expression is a SELECT statement. It is the fundamental building block of most SQL procedure statements. You can combine the results of multiple table expressions with set operators, which creates a query expression. Use one ORDER BY clause for an entire query expression. Place a semicolon only at the end of the entire query expression. A query expression is often only one SELECT statement or table expression.

UPPER Function

Converts the case of a character string to uppercase.

See: "LOWER Function" on page 330

Syntax

UPPER (sql-expression)

Required Argument

sql-expression
must be a character string and is described in "sql-expression" on page 338.

Details

The UPPER function operates on character strings. UPPER converts the case of its argument to all uppercase.

Part 3

Appendixes

Appendix 1

SQL Macro Variables and System Options

Dictionary

SQLCONSTDATETIME System Option

Specifies whether the SQL procedure replaces references to the DATE, TIME, DATETIME, and TODAY functions in a query with their equivalent constant values before the query executes.

Valid in: configuration file, SAS invocation, OPTIONS statement, SAS System Options window

Categories: Files: SAS Files
System administration: SQL

PROC OPTIONS GROUP= SASFILES
SQL

Note: This option can be restricted by a site administrator. For more information, see "Restricted Options" in Chapter 1 of *SAS System Options: Reference*.

Syntax

SQLCONSTDATETIME | NOSQLCONSTDATETIME

Syntax Description

SQLCONSTDATETIME
specifies that the SQL procedure is to replace references to the DATE, TIME, DATETIME, and TODAY functions with their equivalent numeric constant values.

NOSQLCONSTDATETIME
specifies that the SQL procedure is not to replace references to the DATE, TIME, DATETIME, and TODAY functions with their equivalent numeric constant values.

Details

When the SQLCONSTDATETIME system option is set, the SQL procedure evaluates the DATE, TIME, DATETIME, and TODAY functions in a query once, and uses those values throughout the query. Computing these values once ensures consistency of results when the functions are used multiple times in a query or when the query executes the functions close to a date or time boundary.

When the NOSQLCONSTDATETIME system option is set, the SQL procedure evaluates these functions in a query each time it processes an observation.

If both the SQLREDUCEPUT system option and the SQLCONSTDATETIME system option are specified, the SQL procedure replaces the DATE, TIME, DATETIME, and TODAY functions with their respective values in order to determine the PUT function value before the query executes:

```
select x from &lib..c where (put(bday, date9.) = put(today(), date9.));
```

Note: The value that is specified in the SQLCONSTDATETIME system option is in effect for all SQL procedure statements, unless the CONSTDATETIME option in the PROC SQL statement is set. The value of the CONSTDATETIME option takes precedence over the SQLCONSTDATETIME system option. However, changing the value of the CONSTDATETIME option does not change the value of the SQLCONSTDATETIME system option.

See Also

• "Improving Query Performance" on page 140

Procedure Statement Options:

• CONSTDATETIME option on page 217

System Options:

• "SQLREDUCEPUT= System Option" on page 364

SQLGENERATION= System Option

Specifies whether and when SAS procedures generate SQL for in-database processing of source data.

Valid in: configuration file, SAS invocation, OPTIONS statement, SAS System Options window

Default: NONE DBMS='Teradata'

Restriction: For DBMS= and EXCLUDEDB= values, the maximum length of an engine name is 8 characters. For the EXCLUDEPROC= value, the maximum length of a procedure name is 16 characters. An engine can appear only once, and a procedure can appear only once for a given engine.

Data source: Aster *n*Cluster, DB2 under UNIX and PC Hosts, Greenplum, Oracle, Teradata

See: "SQLGENERATION= LIBNAME Option" in *SAS/ACCESS for Relational Databases: Reference* (includes examples), *SAS In-Database Products: User's Guide*

Syntax

SQLGENERATION=<(>NONE | DBMS <DBMS='*engine1 engine2 ... enginen* '>

<EXCLUDEDB=*engine* | '*engine1 engine2 ... enginen*'>

<EXCLUDEPROC="*engine*='*proc1 proc2 ... procn*' *engine2*='*proc1 proc2 ... procn*' *enginen*='*proc1 proc2 ... procn*' "><)>

SQLGENERATION=" "

Syntax Description

NONE
prevents those SAS procedures that are enabled for in-database processing from generating SQL for in database processing. This is a primary state.

DBMS
allows SAS procedures that are enabled for in-database processing to generate SQL for in-database processing of DBMS tables through supported SAS/ACCESS engines. This is a primary state.

DBMS='*engine1 engine2 ... enginen*'
specifies one or more SAS/ACCESS engines. It modifies the primary state.

EXCLUDEDB=*engine* | '*engine1 engine2 ... enginen*'
prevents SAS procedures from generating SQL for in-database processing for one or more specified SAS/ACCESS engines.

EXCLUDEPROC="*engine*='*proc1 proc2 ... procn*' *enginen*='*proc1 proc2 ... procn*' "
identifies engine-specific SAS procedures that do not support in-database processing.

" "
resets the value to the default that was shipped.

Details

Use this option with such procedures as PROC FREQ to indicate what SQL is generated for in-database processing based on the type of subsetting that you need and the SAS/ACCESS engines that you want to access the source table.

You must specify NONE and DBMS, which indicate the primary state.

The maximum length of the option value is 4096. Also, parentheses are required when this option value contains multiple keywords.

Not all procedures support SQL generation for in-database processing for every engine type. If you specify a setting that is not supported, an error message indicates the level of SQL generation that is not supported, and the procedure can reset to the default so that source table records can be read and processed within SAS. If this is not possible, the procedure ends and sets SYSERR= as needed.

You can specify different SQLGENERATION= values for the DATA= and OUT= data sets by using different LIBNAME statements for each of these data sets.

Here is how SAS/ACCESS handles precedence.

Table A1.1 *Precedence of Values for SQLGENERATION= LIBNAME and System Options*

LIBNAME Option	PROC EXCLUDE on System Option?	Engine Type	Engine Specified on System Option	Resulting Value	From (option)
not set NONE DBMS	yes	database interface	NONE DBMS	NONE EXCLUDEDB	system
NONE	no			NONE	LIBNAME
DBMS				DBMS	
not set			NONE	NONE	system
NONE DBMS			DBMS	DBMS	LIBNAME
		no SQL generated for this database host or database version	NONE DBMS	NONE	
not set		Base			system
NONE DBMS					LIBNAME

Example

Here is the default that is shipped with the product.

```
options sqlgeneration='' ;
proc options option=sqlgeneration
 run;
```

SAS procedures generate SQL for in-database processing for all databases except DB2 in this example.

```
options sqlgeneration='' ;
options sqlgeneration=(DBMS EXCLUDEDB='DB2') ;
proc options option=sqlgeneration ;
run;
```

In this example, in-database processing occurs only for Teradata, but SAS procedures generate no SQL for in-database processing.

```
options sqlgeneration='' ;
options SQLGENERATION = (NONE DBMS='Teradata') ;
proc options option=sqlgeneration ;
run;
```

In this next example, SAS procedures do not generate SQL for in-database processing even though in-database processing occurs only for Teradata.

```
options sqlgeneration='' ;
Options SQLGENERATION = (NONE DBMS='Teradata' EXCLUDEDB='DB2') ;
proc options option=sqlgeneration ;
run;
```

For this example, PROC1 and PROC2 for Oracle do not support in-database processing, SAS procedures for Oracle that support in-database processing do not generate SQL for in-database processing, and in-database processing occurs only for Teradata.

```
options sqlgeneration='' ;
Options SQLGENERATION = (NONE EXCLUDEPROC="oracle='proc1,proc2'"
        DBMS='Teradata' EXCLUDEDB='ORACLE') ;
proc options option=sqlgeneration ;
run;
```

SQLMAPPUTTO= System Option

Specifies whether the PUT function is mapped to the SAS_PUT() function for a database, possible also where the SAS_PUT() function is mapped.

Valid in: configuration file, SAS invocation, OPTIONS statement

Default: SAS_PUT

Data source: DB2 under UNIX, Netezza, Teradata

See: "SQL_FUNCTIONS= LIBNAME Option" in *SAS/ACCESS for Relational Databases: Reference*, *SAS In-Database Products: User's Guide*

Syntax

SQLMAPPUTTO= NONE | SAS_PUT | (*database*.SAS_PUT)

Syntax Description

NONE
specifies to PROC SQL that no PUT mapping is to occur.

SAS_PUT
specifies that the PUT function be mapped to the SAS_PUT() function.

***database*.SAS_PUT**
specifies the database name.

TIP It is not necessary that the format definitions and the SAS_PUT() function reside in the same database as the one that contains the data that you want to format. You can use the *database*.SAS_PUT argument to specify the database where the format definitions and the SAS_PUT() function have been published.

TIP The database name can be a multilevel name and it can include blanks.

Requirement: If you specify a database name, you must enclose the entire argument in parentheses.

Details

The format publishing macros deploy or publish, the PUT function implementation to the database as a new function named SAS_PUT(). The format publishing macros also publish both user-defined formats and formats that SAS supplies that you create using

PROC FORMAT. The SAS_PUT() function supports the use of SAS formats, and you can use it in SQL queries that SAS submits to the database so that the entire SQL query can be processed inside the database. You can also use it in conjunction with in-database procedures.

You can use this option with the SQLREDUCEPUT=, SQLREDUCEPUTOBS, and SQLREDUCEPUTVALUES= system options. For more information about these options, see *SAS SQL Procedure User's Guide*.

SQLREDUCEPUT= System Option

For the SQL procedure, specifies the engine type to use to optimize a PUT function in a query. The PUT function is replaced with a logically equivalent expression.

Valid in: configuration file, SAS invocation, OPTIONS statement, SAS System Options window

Categories: Files: SAS Files
System administration: SQL
System administration: Performance

PROC OPTIONS GROUP= SASFILES
SQL
PERFORMANCE

Note: This option can be restricted by a site administrator. For more information, see "Restricted Options" in Chapter 1 of *SAS System Options: Reference*.

Syntax

SQLREDUCEPUT= ALL | NONE | DBMS | BASE

Syntax Description

ALL
specifies to consider the optimization of all PUT functions, regardless of the engine that is used by the query to access the data.

NONE
specifies to not optimize any PUT function.

DBMS
specifies to consider the optimization of all PUT functions in a query performed by a SAS/ACCESS engine. This is the default.

Requirement: The first argument to the PUT function must be a variable that is obtained by a table. The table must be accessed using a SAS/ACCESS engine.

BASE
specifies to consider the optimization of all PUT functions in a query performed by a SAS/ACCESS engine or a Base SAS engine.

Details

If you specify the SQLREDUCEPUT= system option, SAS optimizes the PUT function before the query is executed. If the query also contains a WHERE clause, the evaluation of the WHERE clause is simplified. The following SELECT statements are examples of queries that are optimized if the SQLREDUCEPUT= option is set to any value other than **none**:

```
select x, y from &lib..b where (PUT(x, abc.) in ('yes', 'no'));
select x from &lib..a where (PUT(x, udfmt.) = trim(left('small')));
```

If both the SQLREDUCEPUT= system option and the SQLCONSTDATETIME system option are specified, PROC SQL replaces the DATE, TIME, DATETIME, and TODAY functions with their respective values to determine the PUT function value before the query executes.

The following two SELECT clauses show the original query and optimized query:

```
select x from &lib..c where (put(bday, date9.) = put(today(), date9.));
```

Here, the SELECT clause is optimized.

```
select x from &lib..c where (x = '17MAR2011'D);
```

If a query does not contain the PUT function, it is not optimized.

Note: The value that is specified in the SQLREDUCEPUT= system option is in effect for all SQL procedure statements, unless the PROC SQL REDUCEPUT= option is set. The value of the REDUCEPUT= option takes precedence over the SQLREDUCEPUT= system option. However, changing the value of the REDUCEPUT= option does not change the value of the SQLREDUCEPUT= system option.

See Also

- "Improving Query Performance" on page 140

Procedure Statement Options:

- REDUCEPUT= option on page 220

System Options:

- "SQLCONSTDATETIME System Option" on page 359
- "SQLREDUCEPUTOBS= System Option" on page 365

SQLREDUCEPUTOBS= System Option

For the SQL procedure, when the SQLREDUCEPUT= system option is set to DBMS, BASE, or ALL, specifies the minimum number of observations that must be in a table for PROC SQL to optimize the PUT function in a query.

Valid in: configuration file, SAS invocation, OPTIONS statement, SAS System Options window

Categories: Files: SAS Files
System administration: SQL
System administration: Performance

PROC OPTIONS GROUP= SASFILES
SQL
PERFORMANCE

Interactions: If the SQLREDUCEPUT= system option is set to DBMS, BASE, or ALL, conditions for both the SQLREDUCEPUTOBS= and SQLREDUCEPUTVALUES= system options must be met for PROC SQL to optimize the PUT function.

The SQLREDUCEPUTOBS= system option works only for DBMSs that record the number of observations in a table. If your DBMS does not record the number of observations, but you create row counts on your table, the SQLREDUCEPUTOBS= option will work.

Note: This option can be restricted by a site administrator. For more information, see “Restricted Options” in Chapter 1 of *SAS System Options: Reference*.

Syntax

SQLREDUCEPUTOBS= *n*

Syntax Description

n

specifies the minimum number of observations that must be in a table for PROC SQL to optimize the PUT function in a query.

Default: 0, which indicates that there is no minimum number of observations in a table for PROC SQL to optimize the PUT function.

Range: $0–2^{63}–1$, or approximately 9.2 quintillion

Requirement: *n* must be an integer

Details

For databases that allow implicit pass-through when the row count for a table is not known, PROC SQL allows the PUT function to be optimized in the query, and the query is executed by the database. When the SQLREDUCEPUT= system option is set to DBMS, BASE, or ALL, PROC SQL considers the values of both the SQLREDUCEPUTVALUES= and SQLREDUCEPUTOBS= system options, and determines whether to optimize the PUT function.

For databases that do not allow implicit pass-through, PROC SQL does not optimize the PUT function, and more of the query is executed by SAS.

See Also

- “Improving Query Performance” on page 140

System Options:

- “SQLREDUCEPUT= System Option” on page 364
- “SQLREDUCEPUTVALUES= System Option” on page 366

SQLREDUCEPUTVALUES= System Option

For the SQL procedure, when the SQLREDUCEPUT= system option is set to DBMS, BASE, or ALL, specifies the maximum number of SAS format values that can exist in a PUT function expression for PROC SQL to optimize the PUT function in a query.

Valid in: configuration file, SAS invocation, OPTIONS statement, SAS System Options window

Categories: Files: SAS Files
System administration: SQL
System administration: Performance

PROC OPTIONS GROUP= SASFILES
SQL
PERFORMANCE

Interaction: If the SQLREDUCEPUT= system option is set to DBMS, BASE, or ALL, conditions for both the SQLREDUCEPUTVALUES= and SQLREDUCEPUTOBS= system options must be met for PROC SQL to optimize the PUT function.

Note: This option can be restricted by a site administrator. For more information, see "Restricted Options" in Chapter 1 of *SAS System Options: Reference*.

Syntax

SQLREDUCEPUTVALUES= *n*

Syntax Description

n

specifies the maximum number of SAS format values that can exist in a PUT function expression for PROC SQL to optimize the PUT function in a query.

Default: 100

Range: 100 – 3,000

Requirement: *n* must be an integer

Interaction: If the number of SAS format values in a PUT function expression is greater than this value, PROC SQL does not optimize the PUT function.

Details

Some formats, especially user-defined formats, can contain many format values. Depending on the number of matches for a PUT function expression, the resulting expression can list many format values. If the number of format values becomes too large, query performance can degrade. When the SQLREDUCEPUT= system option is set to DBMS, BASE, or ALL, PROC SQL considers the values of both the SQLREDUCEPUTVALUES= and SQLREDUCEPUTOBS= system options, and determines whether to optimize the PUT function.

TIP The value for SQLREDUCEPUTVALUES= is used for each individual optimization. For example, if you have a PUT function in a WHERE clause, and another PUT function in a GROUP BY clause, the value of SQLREDUCEPUTVALUES= is applied separately for each clause.

See Also

- "Improving Query Performance" on page 140

System Options:

- "SQLREDUCEPUT= System Option" on page 364
- "SQLREDUCEPUTOBS= System Option" on page 365

SQLREMERGE System Option

Specifies whether PROC SQL can process queries that use remerged data.

Valid in: configuration file, SAS invocation, OPTIONS statement, SAS System Options window

Categories: Files: SAS Files
System administration: SQL

PROC OPTIONS GROUP= SASFILES
SQL

Note: This option can be restricted by a site administrator. For more information, see "Restricted Options" in Chapter 1 of *SAS System Options: Reference*.

Syntax

SQLREMERGE | NOSQLREMERGE

Syntax Description

SQLREMERGE
specifies that PROC SQL can process queries that use remerged data.

NOSQLREMERGE
specifies that PROC SQL cannot process queries that use remerged data.

Details

The remerge feature of PROC SQL makes two passes through a table. Data that is created in the first pass is used in the second pass to complete a query. When the NOSQLREMERGE system option is specified, PROC SQL cannot process this remerging of data. If remerging is attempted when the NOSQLREMERGE system option is specified, an error is written to the SAS log.

See Also

- "Improving Query Performance" on page 140

Procedure Statement Options:

- REMERGE option on page 222
- "summary-function" on page 347

SQLUNDOPOLICY= System Option

Specifies how PROC SQL handles updated data if errors occur while you are updating data. You can use UNDO_POLICY= to control whether your changes are permanent.

Valid in: configuration file, SAS invocation, Options statement

Categories: Files: SAS Files

System administration: SQL

PROC OPTIONS GROUP= SASFILES
SQL

Note: This option can be restricted by a site administrator. For more information, see "Restricted Options" in Chapter 1 of *SAS System Options: Reference*.

Syntax

SQLUNDOPOLICY=NONE | OPTIONAL | REQUIRED

Syntax Description

NONE
keeps any updates or inserts.

OPTIONAL
reverses any updates or inserts that it can reverse reliably.

REQUIRED
reverses all inserts or updates that have been done to the point of the error. This is the default.

CAUTION: Some UNDO operations cannot be done reliably. In some cases, the UNDO operation cannot be done reliably. When a change cannot be reversed, PROC SQL issues an error message and does not execute the statement. For example, when a program uses a SAS/ACCESS view, or when a SAS data set is accessed through a SAS/SHARE server and is opened with the data set option CNTLLEV=RECORD, you cannot reliably reverse your changes.

CAUTION: Some UNDO operations might not reverse changes. When multiple transactions are made to the same record, PROC SQL might not reverse a change. PROC SQL issues an error message instead. For example, if an error occurs during an insert, PROC SQL can delete a record that another user updated. In that case, the UNDO operation does not reverse the change, and an error message is issued.

Details

The value that is specified in the SQLUNDOPOLICY= system option is in effect for all SQL procedure statements, unless the PROC SQL UNDO_POLICY= option is set. The value of the UNDO_POLICY= option takes precedence over the SQLUNDOPOLICY= system option. The RESET statement can also be used to set or reset the UNDO_POLICY= option. However, changing the value of the UNDO_POLICY= option does not change the value of the SQLUNDOPOLICY= system option. After the procedure completes, it reverts to the value of the SQLUNDOPOLICY= system option.

If you are updating a data set using the SAS Scalable Performance Data Engine, you can significantly improve processing performance by setting SQLUNDOPOLICY=NONE. However, ensure that NONE is an appropriate setting for your application.

See Also

Procedure Statement

- UNDO_POLICY on page 223

SYS_SQLSETLIMIT Macro Variable

For the SQL procedure, specifies the maximum number of values that is used to optimize a hash join during DBMS processing.

Syntax

SYS_SQLSETLIMIT= *n*;

Required Argument

n

specifies the maximum number of values in the IN condition that is passed to the DBMS for processing,

Default: 1024

Restriction: The SYS_SQLSETLIMIT macro variable affects only certain hash joins.

Example:

```
%let SYS_SQLSETLIMIT=250;
   %let SYS_SQLSETLIMIT=1200;
```

Details

Hash Join

To optimize performance, the SQL procedure might use a hash join when an index join is eliminated as a possibility. With a hash join, the smaller table is reconfigured in memory as a hash table. PROC SQL sequentially scans the larger table, and performs a row-by-row hash lookup against the small table to form the result set. A memory-sizing formula determines whether a hash join is used. The formula is based on the PROC SQL BUFFERSIZE option, whose default value is 64 KB. On a memory-rich system, you should consider increasing BUFFERSIZE to increase the likelihood that a hash join is used.

Appendix 2
PROC SQL and the ANSI Standard

Compliance

PROC SQL follows most of the guidelines set by the American National Standards Institute (ANSI) in its implementation of SQL. However, it is not fully compliant with the current ANSI standard for SQL.[2]

The SQL research project at SAS has focused primarily on the expressive power of SQL as a query language. Consequently, some of the database features of SQL have not yet been implemented in PROC SQL.

SQL Procedure Enhancements

Reserved Words

PROC SQL reserves very few keywords, and then, only in certain contexts. The ANSI standard reserves all SQL keywords in all contexts. For example, according to the standard, you cannot name a column GROUP because of the keywords GROUP BY.

The following words are reserved in PROC SQL:

- The keyword CASE is always reserved. Its use in the CASE expression (an SQL2 feature) precludes its use as a column name.

  If you have a column named CASE in a table, and you want to specify it in a PROC SQL step, then you can use the SAS data set option RENAME= to rename that column for the duration of the query. You can enclose CASE in double quotation marks ("CASE"), and set the PROC SQL option DQUOTE=ANSI.

- The keywords AS, ON, FULL, JOIN, LEFT, FROM, WHEN, WHERE, ORDER, GROUP, RIGHT, INNER, OUTER, UNION, EXCEPT, HAVING, and INTERSECT cannot be used for table aliases. These keywords introduce clauses that appear after a table name. Because the table alias is optional, PROC SQL handles this ambiguity by assuming that any one of these words introduces the corresponding clause and is not the table alias. If you want to use one of these keywords as a table alias, then enclose the keyword in double quotation marks, and set the PROC SQL option DQUOTE=ANSI.

- The keyword USER is reserved for the current user ID. If you specify USER in a SELECT statement in conjunction with a CREATE TABLE statement, then the column is created in the table with a temporary column name that is similar to _TEMA001. If you specify USER in a SELECT statement without a CREATE TABLE statement, then the column is written to the output without a column heading. In either case, the value for the column varies by operating environment,

2 International Organization for Standardization (ISO): Database SQL. Document ISO/IEC 9075:1992. Also available as American National Standards Institute (ANSI) Document ANSI X3.135-1992.

but is typically the user ID of the user who is submitting the program, or the value of the &SYSJOBID automatic macro variable.

If you have a column named USER in a table, and you want to specify it in a PROC SQL step, then you can use the SAS data set option RENAME= to rename that column for the duration of the query. You can enclose USER in double quotation marks ("USER"), and set the PROC SQL option DQUOTE=ANSI.

Column Modifiers

PROC SQL supports the SAS INFORMAT=, FORMAT=, and LABEL= modifiers for expressions in the SELECT statement. These modifiers control the format in which output data is displayed and labeled.

Alternate Collating Sequences

PROC SQL enables you to specify an alternate collating (sorting) sequence to be used when you specify the ORDER BY clause. For more information about the SORTSEQ= option, see PROC SQL Statement on page 215.

ORDER BY Clause in a View Definition

PROC SQL permits you to specify an ORDER BY clause in a CREATE VIEW statement. When the view is queried, its data is sorted based on the specified order, unless a query against that view includes a different ORDER BY clause. For more information, see CREATE VIEW Statement on page 234.

CONTAINS Condition

PROC SQL enables you to test whether a string is part of a column's value when you specify the CONTAINS condition. For more information, see "CONTAINS Condition" on page 312.

Inline Views

The ability to code nested query expressions in the FROM clause is a requirement of the ANSI standard. PROC SQL supports nested coding.

Outer Joins

The ability to include columns that both match and do not match in a join expression is a requirement of the ANSI standard. PROC SQL supports this ability.

Arithmetic Operators

PROC SQL supports the SAS exponentiation (**) operator. PROC SQL uses the notation <> to mean not equal.

Orthogonal Expressions

PROC SQL enables the combination of comparison, Boolean, and algebraic expressions. For example, (X=3)*7 yields a value of 7 if X=3 is true because true is defined to be 1. If X=3 is false, then it resolves to 0, and the entire expression yields a value of 0.

PROC SQL permits a subquery in any expression. This feature is required by the ANSI standard. Therefore, you can have a subquery on the left side of a comparison operator in the WHERE expression.

PROC SQL permits you to order and group data by any type of mathematical expression (except a mathematical expression including a summary function) using ORDER BY and GROUP BY clauses. You can group by an expression that appears in the SELECT statement by using the integer that represents the expression's ordinal position in the SELECT statement. You are not required to select the expression by which you are grouping or ordering. For more information, see "ORDER BY Clause" on page 301 and "GROUP BY Clause" on page 299.

Set Operators

The set operators UNION, INTERSECT, and EXCEPT are required by the ANSI standard. PROC SQL provides these operators and the OUTER UNION operator.

The ANSI standard requires that the tables being operated on have the same number of columns with matching data types. The SQL procedure works on tables that have the same number of columns, and it works on tables that have a different number of columns by creating virtual columns so that a query can evaluate correctly. For more information, see "query-expression" on page 330.

Statistical Functions

PROC SQL supports many more summary functions than required by the ANSI standard for SQL.

PROC SQL supports remerging summary function results into the table's original data. For example, computing the percentage of total is achieved with `100*x/SUM(x)` in PROC SQL. For more information about summary functions and remerging data, see "summary-function" on page 347.

SAS DATA Step Functions

PROC SQL supports many of the functions available in the SAS DATA step. Some of the functions that are not supported are the variable information functions and functions that work with arrays of data. Other SQL databases support their own sets of functions.

PROC FCMP Functions

PROC SQL supports any user-written functions, except those functions with array elements that are created using Chapter 18, "FCMP Procedure" in *Base SAS Procedures Guide*.

SQL Procedure Omissions

COMMIT Statement

The COMMIT statement is not supported.

ROLLBACK Statement

The ROLLBACK statement is not supported. The PROC SQL UNDO_POLICY= option or the SQLUNDOPOLICY system option addresses rollback. See the description of the UNDO_POLICY= option in PROC SQL Statement on page 215 or in the"SQLUNDOPOLICY= System Option" on page 368.

Identifiers and Naming Conventions

In SAS, table names, column names, and aliases are limited to 32 characters, and can contain mixed case. For more information about SAS naming conventions, see *Base SAS Utilities: Reference*. The ANSI standard for SQL allows longer names.

Granting User Privileges

The GRANT statement, PRIVILEGES keyword, and authorization-identifier features of SQL are not supported. You might want to use operating-environment-specific means of security instead.

Three-Valued Logic

ANSI-compatible SQL has three-valued logic. That is, it has special cases for handling comparisons involving NULL values. Any value compared with a NULL value evaluates to NULL.

PROC SQL follows the SAS convention for handling missing values. When numeric NULL values are compared with non-NULL numbers, the NULL values are less than or smaller than all the non-NULL values. When character NULL values are compared with non-NULL characters, the character NULL values are treated as a string of blanks.

Embedded SQL

Currently, there is no provision for embedding PROC SQL statements in other SAS programming environments, such as the DATA step or SAS/IML software.

Appendix 3
Source for SQL Examples

Overview

This section provides the DATA steps to create the tables used in the PROC SQL examples in this guide.

EMPLOYEES

```
data Employees;
   input IdNum $4. +2 LName $11. FName $11. JobCode $3.
         +1 Salary 5. +1 Phone $12.;
   datalines;
```

```
1876  CHIN       JACK       TA1 42400 212/588-5634
1114  GREENWALD  JANICE     ME3 38000 212/588-1092
1556  PENNINGTON MICHAEL    ME1 29860 718/383-5681
1354  PARKER     MARY       FA3 65800 914/455-2337
1130  WOOD       DEBORAH    PT2 36514 212/587-0013
;
```

HOUSES

```
data houses;
   input House $ x y;
   datalines;
house1 1 1
house2 3 3
house3 2 3
house4 7 7
;
```

MATCH_11

```
data match_11;
   input Pair Low Age Lwt Race Smoke Ptd Ht UI @@;
   select(race);
      when (1) do;
         race1=0;
         race2=0;
      end;
      when (2) do;
         race1=1;
         race2=0;
      end;
      when (3) do;
         race1=0;
         race2=1;
      end;
   end;
   datalines;
1  0 14 135 1 0 0 0 0     1  1 14 101 3 1 1 0 0
2  0 15  98 2 0 0 0 0     2  1 15 115 3 0 0 0 1
3  0 16  95 3 0 0 0 0     3  1 16 130 3 0 0 0 0
4  0 17 103 3 0 0 0 0     4  1 17 130 3 1 1 0 1
5  0 17 122 1 1 0 0 0     5  1 17 110 1 1 0 0 0
6  0 17 113 2 0 0 0 0     6  1 17 120 1 1 0 0 0
7  0 17 113 2 0 0 0 0     7  1 17 120 2 0 0 0 0
8  0 17 119 3 0 0 0 0     8  1 17 142 2 0 0 1 0
9  0 18 100 1 1 0 0 0     9  1 18 148 3 0 0 0 0
10 0 18  90 1 1 0 0 1     10 1 18 110 2 1 1 0 0
11 0 19 150 3 0 0 0 0     11 1 19  91 1 1 1 0 1
12 0 19 115 3 0 0 0 0     12 1 19 102 1 0 0 0 0
13 0 19 235 1 1 0 1 0     13 1 19 112 1 1 0 0 1
```

```
14 0 20 120 3 0 0 0 1     14 1 20 150 1 1 0 0 0
15 0 20 103 3 0 0 0 0     15 1 20 125 3 0 0 0 1
16 0 20 169 3 0 1 0 1     16 1 20 120 2 1 0 0 0
17 0 20 141 1 0 1 0 1     17 1 20  80 3 1 0 0 1
18 0 20 121 2 1 0 0 0     18 1 20 109 3 0 0 0 0
19 0 20 127 3 0 0 0 0     19 1 20 121 1 1 1 0 1
20 0 20 120 3 0 0 0 0     20 1 20 122 2 1 0 0 0
21 0 20 158 1 0 0 0 0     21 1 20 105 3 0 0 0 0
22 0 21 108 1 1 0 0 1     22 1 21 165 1 1 0 1 0
23 0 21 124 3 0 0 0 0     23 1 21 200 2 0 0 0 0
24 0 21 185 2 1 0 0 0     24 1 21 103 3 0 0 0 0
25 0 21 160 1 0 0 0 0     25 1 21 100 3 0 1 0 0
26 0 21 115 1 0 0 0 0     26 1 21 130 1 1 0 1 0
27 0 22  95 3 0 0 1 0     27 1 22 130 1 1 0 0 0
28 0 22 158 2 0 1 0 0     28 1 22 130 1 1 1 0 1
29 0 23 130 2 0 0 0 0     29 1 23  97 3 0 0 0 1
30 0 23 128 3 0 0 0 0     30 1 23 187 2 1 0 0 0
31 0 23 119 3 0 0 0 0     31 1 23 120 3 0 0 0 0
32 0 23 115 3 1 0 0 0     32 1 23 110 1 1 1 0 0
33 0 23 190 1 0 0 0 0     33 1 23  94 3 1 0 0 0
34 0 24  90 1 1 1 0 0     34 1 24 128 2 0 1 0 0
35 0 24 115 1 0 0 0 0     35 1 24 132 3 0 0 1 0
36 0 24 110 3 0 0 0 0     36 1 24 155 1 1 1 0 0
37 0 24 115 3 0 0 0 0     37 1 24 138 1 0 0 0 0
38 0 24 110 3 0 1 0 0     38 1 24 105 2 1 0 0 0
39 0 25 118 1 1 0 0 0     39 1 25 105 3 0 1 1 0
40 0 25 120 3 0 0 0 1     40 1 25  85 3 0 0 0 1
41 0 25 155 1 0 0 0 0     41 1 25 115 3 0 0 0 0
42 0 25 125 2 0 0 0 0     42 1 25  92 1 1 0 0 0
43 0 25 140 1 0 0 0 0     43 1 25  89 3 0 1 0 0
44 0 25 241 2 0 0 1 0     44 1 25 105 3 0 1 0 0
45 0 26 113 1 1 0 0 0     45 1 26 117 1 1 1 0 0
46 0 26 168 2 1 0 0 0     46 1 26  96 3 0 0 0 0
47 0 26 133 3 1 1 0 0     47 1 26 154 3 0 1 1 0
48 0 26 160 3 0 0 0 0     48 1 26 190 1 1 0 0 0
49 0 27 124 1 1 0 0 0     49 1 27 130 2 0 0 0 1
50 0 28 120 3 0 0 0 0     50 1 28 120 3 1 1 0 1
51 0 28 130 3 0 0 0 0     51 1 28  95 1 1 0 0 0
52 0 29 135 1 0 0 0 0     52 1 29 130 1 0 0 0 1
53 0 30  95 1 1 0 0 0     53 1 30 142 1 1 1 0 0
54 0 31 215 1 1 0 0 0     54 1 31 102 1 1 1 0 0
55 0 32 121 3 0 0 0 0     55 1 32 105 1 1 0 0 0
56 0 34 170 1 0 1 0 0     56 1 34 187 2 1 0 1 0
;
```

PROCLIB.DELAY

```
data proclib.delay;
   input flight $3. +5 date date7. +2 orig $3. +3 dest $3. +3
         delaycat $15. +2 destype $15. +8 delay;
   informat date date7.;
   format date date7.;
   datalines;
```

```
114    01MAR08  LGA  LAX  1-10 Minutes   Domestic        8
202    01MAR08  LGA  ORD  No Delay       Domestic       -5
219    01MAR08  LGA  LON  11+ Minutes    International  18
622    01MAR08  LGA  FRA  No Delay       International  -5
132    01MAR08  LGA  YYZ  11+ Minutes    International  14
271    01MAR08  LGA  PAR  1-10 Minutes   International   5
302    01MAR08  LGA  WAS  No Delay       Domestic       -2
114    02MAR08  LGA  LAX  No Delay       Domestic        0
202    02MAR08  LGA  ORD  1-10 Minutes   Domestic        5
219    02MAR08  LGA  LON  11+ Minutes    International  18
622    02MAR08  LGA  FRA  No Delay       International   0
132    02MAR08  LGA  YYZ  1-10 Minutes   International   5
271    02MAR08  LGA  PAR  1-10 Minutes   International   4
302    02MAR08  LGA  WAS  No Delay       Domestic        0
114    03MAR08  LGA  LAX  No Delay       Domestic       -1
202    03MAR08  LGA  ORD  No Delay       Domestic       -1
219    03MAR08  LGA  LON  1-10 Minutes   International   4
622    03MAR08  LGA  FRA  No Delay       International  -2
132    03MAR08  LGA  YYZ  1-10 Minutes   International   6
271    03MAR08  LGA  PAR  1-10 Minutes   International   2
302    03MAR08  LGA  WAS  1-10 Minutes   Domestic        5
114    04MAR08  LGA  LAX  11+ Minutes    Domestic       15
202    04MAR08  LGA  ORD  No Delay       Domestic       -5
219    04MAR08  LGA  LON  1-10 Minutes   International   3
622    04MAR08  LGA  FRA  11+ Minutes    International  30
132    04MAR08  LGA  YYZ  No Delay       International  -5
271    04MAR08  LGA  PAR  1-10 Minutes   International   5
302    04MAR08  LGA  WAS  1-10 Minutes   Domestic        7
114    05MAR08  LGA  LAX  No Delay       Domestic       -2
202    05MAR08  LGA  ORD  1-10 Minutes   Domestic        2
219    05MAR08  LGA  LON  1-10 Minutes   International   3
622    05MAR08  LGA  FRA  No Delay       International  -6
132    05MAR08  LGA  YYZ  1-10 Minutes   International   3
271    05MAR08  LGA  PAR  1-10 Minutes   International   5
114    06MAR08  LGA  LAX  No Delay       Domestic       -1
202    06MAR08  LGA  ORD  No Delay       Domestic       -3
219    06MAR08  LGA  LON  11+ Minutes    International  27
132    06MAR08  LGA  YYZ  1-10 Minutes   International   7
302    06MAR08  LGA  WAS  1-10 Minutes   Domestic        1
114    07MAR08  LGA  LAX  No Delay       Domestic       -1
202    07MAR08  LGA  ORD  No Delay       Domestic       -2
219    07MAR08  LGA  LON  11+ Minutes    International  15
622    07MAR08  LGA  FRA  11+ Minutes    International  21
132    07MAR08  LGA  YYZ  No Delay       International  -2
271    07MAR08  LGA  PAR  1-10 Minutes   International   4
302    07MAR08  LGA  WAS  No Delay       Domestic        0
;
```

PROCLIB.HOUSES

The contents of this data set are different from the "HOUSES" on page 376 data set. This data set is intended only for the "Example: INTO Clause" on page 292.

```
libname proclib 'SAS-library';

data proclib.houses;
input Style $ 1-8 SqFeet 15-18;
datalines;
CONDO          900
CONDO         1000
RANCH         1200
RANCH         1400
SPLIT         1600
SPLIT         1800
TWOSTORY      2100
TWOSTORY      3000
TWOSTORY      1940
TWOSTORY      1860
;
```

PROCLIB.MARCH

```
data proclib.march;
   input flight $3. +5 date date7. +3 depart time5. +2 orig $3.
         +3 dest $3.  +7 miles +6 boarded +6 capacity;
   format date date7. depart time5.;
   informat date date7. depart time5.;
   datalines;
114     01MAR08    7:10  LGA   LAX        2475        172         210
202     01MAR08   10:43  LGA   ORD         740        151         210
219     01MAR08    9:31  LGA   LON        3442        198         250
622     01MAR08   12:19  LGA   FRA        3857        207         250
132     01MAR08   15:35  LGA   YYZ         366        115         178
271     01MAR08   13:17  LGA   PAR        3635        138         250
302     01MAR08   20:22  LGA   WAS         229        105         180
114     02MAR08    7:10  LGA   LAX        2475        119         210
202     02MAR08   10:43  LGA   ORD         740        120         210
219     02MAR08    9:31  LGA   LON        3442        147         250
622     02MAR08   12:19  LGA   FRA        3857        176         250
132     02MAR08   15:35  LGA   YYZ         366        106         178
302     02MAR08   20:22  LGA   WAS         229         78         180
271     02MAR08   13:17  LGA   PAR        3635        104         250
114     03MAR08    7:10  LGA   LAX        2475        197         210
202     03MAR08   10:43  LGA   ORD         740        118         210
219     03MAR08    9:31  LGA   LON        3442        197         250
622     03MAR08   12:19  LGA   FRA        3857        180         250
132     03MAR08   15:35  LGA   YYZ         366         75         178
271     03MAR08   13:17  LGA   PAR        3635        147         250
302     03MAR08   20:22  LGA   WAS         229        123         180
114     04MAR08    7:10  LGA   LAX        2475        178         210
202     04MAR08   10:43  LGA   ORD         740        148         210
219     04MAR08    9:31  LGA   LON        3442        232         250
622     04MAR08   12:19  LGA   FRA        3857        137         250
132     04MAR08   15:35  LGA   YYZ         366        117         178
271     04MAR08   13:17  LGA   PAR        3635        146         250
```

```
302    04MAR08   20:22  LGA   WAS       229       115       180
114    05MAR08    7:10  LGA   LAX      2475       117       210
202    05MAR08   10:43  LGA   ORD       740       104       210
219    05MAR08    9:31  LGA   LON      3442       160       250
622    05MAR08   12:19  LGA   FRA      3857       185       250
132    05MAR08   15:35  LGA   YYZ       366       157       178
271    05MAR08   13:17  LGA   PAR      3635       177       250
114    06MAR08    7:10  LGA   LAX      2475       128       210
202    06MAR08   10:43  LGA   ORD       740       115       210
219    06MAR08    9:31  LGA   LON      3442       163       250
132    06MAR08   15:35  LGA   YYZ       366       150       178
302    06MAR08   20:22  LGA   WAS       229        66       180
114    07MAR08    7:10  LGA   LAX      2475       160       210
202    07MAR08   10:43  LGA   ORD       740       175       210
219    07MAR08    9:31  LGA   LON      3442       241       250
622    07MAR08   12:19  LGA   FRA      3857       210       250
132    07MAR08   15:35  LGA   YYZ       366       164       178
271    07MAR08   13:17  LGA   PAR      3635       155       250
302    07MAR08   20:22  LGA   WAS       229       135       180
;
```

PROCLIB.PAYLIST2

```
proc sql;
   create table proclib.paylist2
       (IdNum char(4),
        Gender char(1),
        Jobcode char(3),
        Salary num,
        Birth num informat=date7.
                  format=date7.,
        Hired num informat=date7.
                  format=date7.);

insert into proclib.paylist2
values('1919','M','TA2',34376,'12SEP66'd,'04JUN87'd)
values('1653','F','ME2',31896,'15OCT64'd,'09AUG92'd)
values('1350','F','FA3',36886,'31AUG55'd,'29JUL91'd)
values('1401','M','TA3',38822,'13DEC55'd,'17NOV93'd)
values('1499','M','ME1',23025,'26APR74'd,'07JUN92'd);

title 'PROCLIB.PAYLIST2 Table';
select * from proclib.paylist2;
```

PROCLIB.PAYROLL

This data set is updated in “Example 3: Updating Data in a PROC SQL Table” on page 249. Its updated data is used in subsequent examples.

```
data proclib.payroll;
   input IdNumber $4. +3 Gender $1. +4 Jobcode $3. +9 Salary 5.
         +2 Birth date7. +2 Hired date7.;
   informat birth date7. hired date7.;
   format birth date7. hired date7.;
   datalines;
1919    M     TA2          34376  12SEP60  04JUN87
1653    F     ME2          35108  15OCT64  09AUG90
1400    M     ME1          29769  05NOV67  16OCT90
1350    F     FA3          32886  31AUG65  29JUL90
1401    M     TA3          38822  13DEC50  17NOV85
1499    M     ME3          43025  26APR54  07JUN80
1101    M     SCP          18723  06JUN62  01OCT90
1333    M     PT2          88606  30MAR61  10FEB81
1402    M     TA2          32615  17JAN63  02DEC90
1479    F     TA3          38785  22DEC68  05OCT89
1403    M     ME1          28072  28JAN69  21DEC91
1739    M     PT1          66517  25DEC64  27JAN91
1658    M     SCP          17943  08APR67  29FEB92
1428    F     PT1          68767  04APR60  16NOV91
1782    M     ME2          35345  04DEC70  22FEB92
1244    M     ME2          36925  31AUG63  17JAN88
1383    M     BCK          25823  25JAN68  20OCT92
1574    M     FA2          28572  27APR60  20DEC92
1789    M     SCP          18326  25JAN57  11APR78
1404    M     PT2          91376  24FEB53  01JAN80
1437    F     FA3          33104  20SEP60  31AUG84
1639    F     TA3          40260  26JUN57  28JAN84
1269    M     NA1          41690  03MAY72  28NOV92
1065    M     ME2          35090  26JAN44  07JAN87
1876    M     TA3          39675  20MAY58  27APR85
1037    F     TA1          28558  10APR64  13SEP92
1129    F     ME2          34929  08DEC61  17AUG91
1988    M     FA3          32217  30NOV59  18SEP84
1405    M     SCP          18056  05MAR66  26JAN92
1430    F     TA2          32925  28FEB62  27APR87
1983    F     FA3          33419  28FEB62  27APR87
1134    F     TA2          33462  05MAR69  21DEC88
1118    M     PT3         111379  16JAN44  18DEC80
1438    F     TA3          39223  15MAR65  18NOV87
1125    F     FA2          28888  08NOV68  11DEC87
1475    F     FA2          27787  15DEC61  13JUL90
1117    M     TA3          39771  05JUN63  13AUG92
1935    F     NA2          51081  28MAR54  16OCT81
1124    F     FA1          23177  10JUL58  01OCT90
1422    F     FA1          22454  04JUN64  06APR91
1616    F     TA2          34137  01MAR70  04JUN93
1406    M     ME2          35185  08MAR61  17FEB87
1120    M     ME1          28619  11SEP72  07OCT93
1094    M     FA1          22268  02APR70  17APR91
1389    M     BCK          25028  15JUL59  18AUG90
1905    M     PT1          65111  16APR72  29MAY92
1407    M     PT1          68096  23MAR69  18MAR90
1114    F     TA2          32928  18SEP69  27JUN87
1410    M     PT2          84685  03MAY67  07NOV86
```

```
1439   F   PT1     70736  06MAR64  10SEP90
1409   M   ME3     41551  19APR50  22OCT81
1408   M   TA2     34138  29MAR60  14OCT87
1121   M   ME1     29112  26SEP71  07DEC91
1991   F   TA1     27645  07MAY72  12DEC92
1102   M   TA2     34542  01OCT59  15APR91
1356   M   ME2     36869  26SEP57  22FEB83
1545   M   PT1     66130  12AUG59  29MAY90
1292   F   ME2     36691  28OCT64  02JUL89
1440   F   ME2     35757  27SEP62  09APR91
1368   M   FA2     27808  11JUN61  03NOV84
1369   M   TA2     33705  28DEC61  13MAR87
1411   M   FA2     27265  27MAY61  01DEC89
1113   F   FA1     22367  15JAN68  17OCT91
1704   M   BCK     25465  30AUG66  28JUN87
1900   M   ME2     35105  25MAY62  27OCT87
1126   F   TA3     40899  28MAY63  21NOV80
1677   M   BCK     26007  05NOV63  27MAR89
1441   F   FA2     27158  19NOV69  23MAR91
1421   M   TA2     33155  08JAN59  28FEB90
1119   M   TA1     26924  20JUN62  06SEP88
1834   M   BCK     26896  08FEB72  02JUL92
1777   M   PT3    109630  23SEP51  21JUN81
1663   M   BCK     26452  11JAN67  11AUG91
1106   M   PT2     89632  06NOV57  16AUG84
1103   F   FA1     23738  16FEB68  23JUL92
1477   M   FA2     28566  21MAR64  07MAR88
1476   F   TA2     34803  30MAY66  17MAR87
1379   M   ME3     42264  08AUG61  10JUN84
1104   M   SCP     17946  25APR63  10JUN91
1009   M   TA1     28880  02MAR59  26MAR92
1412   M   ME1     27799  18JUN56  05DEC91
1115   F   FA3     32699  22AUG60  29FEB80
1128   F   TA2     32777  23MAY65  20OCT90
1442   F   PT2     84536  05SEP66  12APR88
1417   M   NA2     52270  27JUN64  07MAR89
1478   M   PT2     84203  09AUG59  24OCT90
1673   M   BCK     25477  27FEB70  15JUL91
1839   F   NA1     43433  29NOV70  03JUL93
1347   M   TA3     40079  21SEP67  06SEP84
1423   F   ME2     35773  14MAY68  19AUG90
1200   F   ME1     27816  10JAN71  14AUG92
1970   F   FA1     22615  25SEP64  12MAR91
1521   M   ME3     41526  12APR63  13JUL88
1354   F   SCP     18335  29MAY71  16JUN92
1424   F   FA2     28978  04AUG69  11DEC89
1132   F   FA1     22413  30MAY72  22OCT93
1845   M   BCK     25996  20NOV59  22MAR80
1556   M   PT1     71349  22JUN64  11DEC91
1413   M   FA2     27435  16SEP65  02JAN90
1123   F   TA1     28407  31OCT72  05DEC92
1907   M   TA2     33329  15NOV60  06JUL87
1436   F   TA2     34475  11JUN64  12MAR87
1385   M   ME3     43900  16JAN62  01APR86
1432   F   ME2     35327  03NOV61  10FEB85
1111   M   NA1     40586  14JUL73  31OCT92
```

```
1116   F   FA1        22862  28SEP69  21MAR91
1352   M   NA2        53798  02DEC60  16OCT86
1555   F   FA2        27499  16MAR68  04JUL92
1038   F   TA1        26533  09NOV69  23NOV91
1420   M   ME3        43071  19FEB65  22JUL87
1561   M   TA2        34514  30NOV63  07OCT87
1434   F   FA2        28622  11JUL62  28OCT90
1414   M   FA1        23644  24MAR72  12APR92
1112   M   TA1        26905  29NOV64  07DEC92
1390   M   FA2        27761  19FEB65  23JUN91
1332   M   NA1        42178  17SEP70  04JUN91
1890   M   PT2        91908  20JUL51  25NOV79
1429   F   TA1        27939  28FEB60  07AUG92
1107   M   PT2        89977  09JUN54  10FEB79
1908   F   TA2        32995  10DEC69  23APR90
1830   F   PT2        84471  27MAY57  29JAN83
1882   M   ME3        41538  10JUL57  21NOV78
1050   M   ME2        35167  14JUL63  24AUG86
1425   F   FA1        23979  28DEC71  28FEB93
1928   M   PT2        89858  16SEP54  13JUL90
1480   F   TA3        39583  03SEP57  25MAR81
1100   M   BCK        25004  01DEC60  07MAY88
1995   F   ME1        28810  24AUG73  19SEP93
1135   F   FA2        27321  20SEP60  31MAR90
1415   M   FA2        28278  09MAR58  12FEB88
1076   M   PT1        66558  14OCT55  03OCT91
1426   F   TA2        32991  05DEC66  25JUN90
1564   F   SCP        18833  12APR62  01JUL92
1221   F   FA2        27896  22SEP67  04OCT91
1133   M   TA1        27701  13JUL66  12FEB92
1435   F   TA3        38808  12MAY59  08FEB80
1418   M   ME1        28005  29MAR57  06JAN92
1017   M   TA3        40858  28DEC57  16OCT81
1443   F   NA1        42274  17NOV68  29AUG91
1131   F   TA2        32575  26DEC71  19APR91
1427   F   TA2        34046  31OCT70  30JAN90
1036   F   TA3        39392  19MAY65  23OCT84
1130   F   FA1        23916  16MAY71  05JUN92
1127   F   TA2        33011  09NOV64  07DEC86
1433   F   FA3        32982  08JUL66  17JAN87
1431   F   FA3        33230  09JUN64  05APR88
1122   F   FA2        27956  01MAY63  27NOV88
1105   M   ME2        34805  01MAR62  13AUG90
;
```

PROCLIB.PAYROLL2

```
data proclib.payroll2;
   input idnum $4. +3 gender $1. +4 jobcode $3. +9 salary 5.
         +2 birth date7. +2 hired date7.;
   informat birth date7. hired date7.;
   format birth date7. hired date7.;
```

```
   datalines;
1639  F  TA3      42260  26JUN57  28JAN84
1065  M  ME3      38090  26JAN44  07JAN87
1561  M  TA3      36514  30NOV63  07OCT87
1221  F  FA3      29896  22SEP67  04OCT91
1447  F  FA1      22123  07AUG72  29OCT92
1998  M  SCP      23100  10SEP70  02NOV92
1036  F  TA3      42465  19MAY65  23OCT84
1106  M  PT3      94039  06NOV57  16AUG84
1129  F  ME3      36758  08DEC61  17AUG91
1350  F  FA3      36098  31AUG65  29JUL90
1369  M  TA3      36598  28DEC61  13MAR87
1076  M  PT1      69742  14OCT55  03OCT91
;
```

PROCLIB.SCHEDULE2

```
data proclib.schedule2;
   input flight $3. +5 date date7. +2 dest $3. +3 idnum $4.;
   format date date7.;
   informat date date7.;
   datalines;
132     01MAR94  BOS   1118
132     01MAR94  BOS   1402
219     02MAR94  PAR   1616
219     02MAR94  PAR   1478
622     03MAR94  LON   1430
622     03MAR94  LON   1882
271     04MAR94  NYC   1430
271     04MAR94  NYC   1118
579     05MAR94  RDU   1126
579     05MAR94  RDU   1106
;
```

PROCLIB.STAFF

```
data proclib.staff;
   input idnum $4. +3 lname $15. +2 fname $15. +2 city $15. +2
         state $2. +5 hphone $12.;
   datalines;
1919   ADAMS            GERALD           STAMFORD         CT     203/781-1255
1653   ALIBRANDI        MARIA            BRIDGEPORT       CT     203/675-7715
1400   ALHERTANI        ABDULLAH         NEW YORK         NY     212/586-0808
1350   ALVAREZ          MERCEDES         NEW YORK         NY     718/383-1549
1401   ALVAREZ          CARLOS           PATERSON         NJ     201/732-8787
1499   BAREFOOT         JOSEPH           PRINCETON        NJ     201/812-5665
1101   BAUCOM           WALTER           NEW YORK         NY     212/586-8060
1333   BANADYGA         JUSTIN           STAMFORD         CT     203/781-1777
1402   BLALOCK          RALPH            NEW YORK         NY     718/384-2849
1479   BALLETTI         MARIE            NEW YORK         NY     718/384-8816
```

1403	BOWDEN	EARL	BRIDGEPORT	CT	203/675-3434
1739	BRANCACCIO	JOSEPH	NEW YORK	NY	212/587-1247
1658	BREUHAUS	JEREMY	NEW YORK	NY	212/587-3622
1428	BRADY	CHRISTINE	STAMFORD	CT	203/781-1212
1782	BREWCZAK	JAKOB	STAMFORD	CT	203/781-0019
1244	BUCCI	ANTHONY	NEW YORK	NY	718/383-3334
1383	BURNETTE	THOMAS	NEW YORK	NY	718/384-3569
1574	CAHILL	MARSHALL	NEW YORK	NY	718/383-2338
1789	CARAWAY	DAVIS	NEW YORK	NY	212/587-9000
1404	COHEN	LEE	NEW YORK	NY	718/384-2946
1437	CARTER	DOROTHY	BRIDGEPORT	CT	203/675-4117
1639	CARTER-COHEN	KAREN	STAMFORD	CT	203/781-8839
1269	CASTON	FRANKLIN	STAMFORD	CT	203/781-3335
1065	COPAS	FREDERICO	NEW YORK	NY	718/384-5618
1876	CHIN	JACK	NEW YORK	NY	212/588-5634
1037	CHOW	JANE	STAMFORD	CT	203/781-8868
1129	COUNIHAN	BRENDA	NEW YORK	NY	718/383-2313
1988	COOPER	ANTHONY	NEW YORK	NY	212/587-1228
1405	DACKO	JASON	PATERSON	NJ	201/732-2323
1430	DABROWSKI	SANDRA	BRIDGEPORT	CT	203/675-1647
1983	DEAN	SHARON	NEW YORK	NY	718/384-1647
1134	DELGADO	MARIA	STAMFORD	CT	203/781-1528
1118	DENNIS	ROGER	NEW YORK	NY	718/383-1122
1438	DABBOUSSI	KAMILLA	STAMFORD	CT	203/781-2229
1125	DUNLAP	DONNA	NEW YORK	NY	718/383-2094
1475	ELGES	MARGARETE	NEW YORK	NY	718/383-2828
1117	EDGERTON	JOSHUA	NEW YORK	NY	212/588-1239
1935	FERNANDEZ	KATRINA	BRIDGEPORT	CT	203/675-2962
1124	FIELDS	DIANA	WHITE PLAINS	NY	914/455-2998
1422	FUJIHARA	KYOKO	PRINCETON	NJ	201/812-0902
1616	FUENTAS	CARLA	NEW YORK	NY	718/384-3329
1406	FOSTER	GERALD	BRIDGEPORT	CT	203/675-6363
1120	GARCIA	JACK	NEW YORK	NY	718/384-4930
1094	GOMEZ	ALAN	BRIDGEPORT	CT	203/675-7181
1389	GOLDSTEIN	LEVI	NEW YORK	NY	718/384-9326
1905	GRAHAM	ALVIN	NEW YORK	NY	212/586-8815
1407	GREGORSKI	DANIEL	MT. VERNON	NY	914/468-1616
1114	GREENWALD	JANICE	NEW YORK	NY	212/588-1092
1410	HARRIS	CHARLES	STAMFORD	CT	203/781-0937
1439	HASENHAUER	CHRISTINA	BRIDGEPORT	CT	203/675-4987
1409	HAVELKA	RAYMOND	STAMFORD	CT	203/781-9697
1408	HENDERSON	WILLIAM	PRINCETON	NJ	201/812-4789
1121	HERNANDEZ	ROBERTO	NEW YORK	NY	718/384-3313
1991	HOWARD	GRETCHEN	BRIDGEPORT	CT	203/675-0007
1102	HERMANN	JOACHIM	WHITE PLAINS	NY	914/455-0976
1356	HOWARD	MICHAEL	NEW YORK	NY	212/586-8411
1545	HERRERO	CLYDE	STAMFORD	CT	203/781-1119
1292	HUNTER	HELEN	BRIDGEPORT	CT	203/675-4830
1440	JACKSON	LAURA	STAMFORD	CT	203/781-0088
1368	JEPSEN	RONALD	STAMFORD	CT	203/781-8413
1369	JONSON	ANTHONY	NEW YORK	NY	212/587-5385
1411	JOHNSEN	JACK	PATERSON	NJ	201/732-3678
1113	JOHNSON	LESLIE	NEW YORK	NY	718/383-3003
1704	JONES	NATHAN	NEW YORK	NY	718/384-0049
1900	KING	WILLIAM	NEW YORK	NY	718/383-3698
1126	KIMANI	ANNE	NEW YORK	NY	212/586-1229

1677	KRAMER	JACKSON	BRIDGEPORT	CT	203/675-7432
1441	LAWRENCE	KATHY	PRINCETON	NJ	201/812-3337
1421	LEE	RUSSELL	MT. VERNON	NY	914/468-9143
1119	LI	JEFF	NEW YORK	NY	212/586-2344
1834	LEBLANC	RUSSELL	NEW YORK	NY	718/384-0040
1777	LUFKIN	ROY	NEW YORK	NY	718/383-4413
1663	MARKS	JOHN	NEW YORK	NY	212/587-7742
1106	MARSHBURN	JASPER	STAMFORD	CT	203/781-1457
1103	MCDANIEL	RONDA	NEW YORK	NY	212/586-0013
1477	MEYERS	PRESTON	BRIDGEPORT	CT	203/675-8125
1476	MONROE	JOYCE	STAMFORD	CT	203/781-2837
1379	MORGAN	ALFRED	STAMFORD	CT	203/781-2216
1104	MORGAN	CHRISTOPHER	NEW YORK	NY	718/383-9740
1009	MORGAN	GEORGE	NEW YORK	NY	212/586-7753
1412	MURPHEY	JOHN	PRINCETON	NJ	201/812-4414
1115	MURPHY	ALICE	NEW YORK	NY	718/384-1982
1128	NELSON	FELICIA	BRIDGEPORT	CT	203/675-1166
1442	NEWKIRK	SANDRA	PRINCETON	NJ	201/812-3331
1417	NEWKIRK	WILLIAM	PATERSON	NJ	201/732-6611
1478	NEWTON	JAMES	NEW YORK	NY	212/587-5549
1673	NICHOLLS	HENRY	STAMFORD	CT	203/781-7770
1839	NORRIS	DIANE	NEW YORK	NY	718/384-1767
1347	O'NEAL	BRYAN	NEW YORK	NY	718/384-0230
1423	OSWALD	LESLIE	MT. VERNON	NY	914/468-9171
1200	OVERMAN	MICHELLE	STAMFORD	CT	203/781-1835
1970	PARKER	ANNE	NEW YORK	NY	718/383-3895
1521	PARKER	JAY	NEW YORK	NY	212/587-7603
1354	PARKER	MARY	WHITE PLAINS	NY	914/455-2337
1424	PATTERSON	RENEE	NEW YORK	NY	212/587-8991
1132	PEARCE	CAROL	NEW YORK	NY	718/384-1986
1845	PEARSON	JAMES	NEW YORK	NY	718/384-2311
1556	PENNINGTON	MICHAEL	NEW YORK	NY	718/383-5681
1413	PETERS	RANDALL	PRINCETON	NJ	201/812-2478
1123	PETERSON	SUZANNE	NEW YORK	NY	718/383-0077
1907	PHELPS	WILLIAM	STAMFORD	CT	203/781-1118
1436	PORTER	SUSAN	NEW YORK	NY	718/383-5777
1385	RAYNOR	MILTON	BRIDGEPORT	CT	203/675-2846
1432	REED	MARILYN	MT. VERNON	NY	914/468-5454
1111	RHODES	JEREMY	PRINCETON	NJ	201/812-1837
1116	RICHARDS	CASEY	NEW YORK	NY	212/587-1224
1352	RIVERS	SIMON	NEW YORK	NY	718/383-3345
1555	RODRIGUEZ	JULIA	BRIDGEPORT	CT	203/675-2401
1038	RODRIGUEZ	MARIA	BRIDGEPORT	CT	203/675-2048
1420	ROUSE	JEREMY	PATERSON	NJ	201/732-9834
1561	SANDERS	RAYMOND	NEW YORK	NY	212/588-6615
1434	SANDERSON	EDITH	STAMFORD	CT	203/781-1333
1414	SANDERSON	NATHAN	BRIDGEPORT	CT	203/675-1715
1112	SANYERS	RANDY	NEW YORK	NY	718/384-4895
1390	SMART	JONATHAN	NEW YORK	NY	718/383-1141
1332	STEPHENSON	ADAM	BRIDGEPORT	CT	203/675-1497
1890	STEPHENSON	ROBERT	NEW YORK	NY	718/384-9874
1429	THOMPSON	ALICE	STAMFORD	CT	203/781-3857
1107	THOMPSON	WAYNE	NEW YORK	NY	718/384-3785
1908	TRENTON	MELISSA	NEW YORK	NY	212/586-6262
1830	TRIPP	KATHY	BRIDGEPORT	CT	203/675-2479
1882	TUCKER	ALAN	NEW YORK	NY	718/384-0216

```
1050   TUTTLE         THOMAS         WHITE PLAINS     NY      914/455-2119
1425   UNDERWOOD      JENNY          STAMFORD         CT      203/781-0978
1928   UPCHURCH       LARRY          WHITE PLAINS     NY      914/455-5009
1480   UPDIKE         THERESA        NEW YORK         NY      212/587-8729
1100   VANDEUSEN      RICHARD        NEW YORK         NY      212/586-2531
1995   VARNER         ELIZABETH      NEW YORK         NY      718/384-7113
1135   VEGA           ANNA           NEW YORK         NY      718/384-5913
1415   VEGA           FRANKLIN       NEW YORK         NY      718/384-2823
1076   VENTER         RANDALL        NEW YORK         NY      718/383-2321
1426   VICK           THERESA        PRINCETON        NJ      201/812-2424
1564   WALTERS        ANNE           NEW YORK         NY      212/587-3257
1221   WALTERS        DIANE          NEW YORK         NY      718/384-1918
1133   WANG           CHIN           NEW YORK         NY      212/587-1956
1435   WARD           ELAINE         NEW YORK         NY      718/383-4987
1418   WATSON         BERNARD        NEW YORK         NY      718/383-1298
1017   WELCH          DARIUS         NEW YORK         NY      212/586-5535
1443   WELLS          AGNES          STAMFORD         CT      203/781-5546
1131   WELLS          NADINE         NEW YORK         NY      718/383-1045
1427   WHALEY         CAROLYN        MT. VERNON       NY      914/468-4528
1036   WONG           LESLIE         NEW YORK         NY      212/587-2570
1130   WOOD           DEBORAH        NEW YORK         NY      212/587-0013
1127   WOOD           SANDRA         NEW YORK         NY      212/587-2881
1433   YANCEY         ROBIN          PRINCETON        NJ      201/812-1874
1431   YOUNG          DEBORAH        STAMFORD         CT      203/781-2987
1122   YOUNG          JOANN          NEW YORK         NY      718/384-2021
1105   YOUNG          LAWRENCE       NEW YORK         NY      718/384-0008
;
```

PROCLIB.STAFF2

```
data proclib.staff2;
input IdNum $4. @7 Lname $12. @20 Fname $8. @30 City $10.
      @42 State $2. @50 Hphone $12.;
   datalines;
1106  MARSHBURN    JASPER    STAMFORD    CT       203/781-1457
1430  DABROWSKI    SANDRA    BRIDGEPORT  CT       203/675-1647
1118  DENNIS       ROGER     NEW YORK    NY       718/383-1122
1126  KIMANI       ANNE      NEW YORK    NY       212/586-1229
1402  BLALOCK      RALPH     NEW YORK    NY       718/384-2849
1882  TUCKER       ALAN      NEW YORK    NY       718/384-0216
1479  BALLETTI     MARIE     NEW YORK    NY       718/384-8816
1420  ROUSE        JEREMY    PATERSON    NJ       201/732-9834
1403  BOWDEN       EARL      BRIDGEPORT  CT       203/675-3434
1616  FUENTAS      CARLA     NEW YORK    NY       718/384-3329
;
```

PROCLIB.SUPERV2

```
data proclib.superv2;
   input supid $4. +8 state $2. +5  jobcat  $2.;
   label supid='Supervisor Id' jobcat='Job Category';
```

```
   datalines;
1417        NJ     NA
1352        NY     NA
1106        CT     PT
1442        NJ     PT
1118        NY     PT
1405        NJ     SC
1564        NY     SC
1639        CT     TA
1126        NY     TA
1882        NY     ME
;
```

STORES

```
data stores;
  input Store $ x y;
  datalines;
store1 5 1
store2 5 3
store3 3 5
store4 7 5
;
```

SURVEY

```
data survey;
  input id $ diet $ exer $ hours xwk educ;
  datalines;
1001 yes yes 1 3 1
1002 no  yes 1 4 2
1003 no  no  . . .n
1004 yes yes 2 3 .x
1005 no  yes 2 3 .x
1006 yes yes 2 4 .x
1007 no  yes .5 3 .
1008 no  no  . .  .
;
```

Glossary

calculated column
in a query, a column that does not exist in any of the tables that are being queried, but which is created as a result of a column expression.

Cartesian product
a type of join that matches each row from each joined table to each row from all other joined tables.

column
a vertical component of a table. Each column has a unique name, contains data of a specific type, and has particular attributes. A column is analogous to a variable in SAS terminology.

column alias
a temporary, alternate name for a column. Aliases are optionally specified in the SQL procedure's SELECT clause to name or rename columns. An alias is one word.

column expression
a set of operators and operands that, when evaluated, result in a single data value. The resulting data value can be either a character value or a numeric value.

composite index
an index that locates observations in a SAS data set by examining the values of two or more key variables.

condition
in the SQL procedure, the part of the WHERE clause that specifies which rows are to be retrieved.

cross join
a type of join that returns the product of joined tables. A cross join is functionally the same as a Cartesian product.

data set
See SAS data set.

data view
See SAS data view.

DISTINCT
a keyword that causes the SQL procedure to remove duplicate rows from the output.

equijoin
a kind of join in the SQL procedure. For example, when two tables are joined in an equijoin, the value of a column in the first table must equal the value of the column in the second table in the SQL expression.

format
See SAS format.

group
a set of rows or observations that have the same value or values for one or more common columns or variables.

in-line view
a query-expression that is nested in the SQL procedure's FROM clause. An in-line view produces a table internally that the outer query uses to select data. You save a programming step when you use an in-line view, because instead of creating a view and then referring to it in another query, you can specify the view in-line in the FROM clause. An in-line view can be referenced only in the query (or statement) in which it is defined.

index
a component of a SAS data set that enables SAS to access observations in the SAS data set quickly and efficiently. The purpose of SAS indexes is to optimize WHERE-clause processing and to facilitate BY-group processing.

inner join
a join between two tables that returns all of the rows in one table that have one or more matching rows in the other table.

integrity constraints
a set of data validation rules that you can specify in order to restrict the data values that can be stored for a variable in a SAS data file. Integrity constraints help you preserve the validity and consistency of your data.

join
an operation that combines data from two or more tables. A join is typically created by means of SQL (Structured Query Language) code or a user interface.

join criteria
the set of parameters that determine how tables are to be joined. Join criteria are usually specified in a WHERE expression or in an SQL ON clause.

missing value
a type of value for a variable that contains no data for a particular row or column. By default, SAS writes a missing numeric value as a single period and a missing character value as a blank space.

natural join
a type of join that returns selected rows from tables in which one or more columns in each table have the same name and the same data type and contain the same value.

outer join
a join between two tables that returns all of the rows in one table, as well as part or all of the rows in the other table. A left or right outer join returns all of the rows in one table (the table on the left or right side of the SQL statement, respectively), as

well as the matching rows in the other table. A full outer join returns all of the rows in both of the tables.

pass-through facility
See SQL pass-through facility.

PROC SQL view
a SAS data set that is created by the SQL procedure. A PROC SQL view contains no data. Instead, it stores information that enables it to read data values from other files, which can include SAS data files, SAS/ACCESS views, DATA step views, or other PROC SQL views. The output of a PROC SQL view can be either a subset or a superset of one or more files.

query
a set of instructions that requests particular information from one or more data sources.

query expression
in PROC SQL, a SELECT statement that references at least one table and, when executed, creates a temporary table that exists only during the execution of the statement. You can combine the results of multiple table expressions with set operators to create a query expression.

SAS data file
a type of SAS data set that contains data values as well as descriptor information that is associated with the data. The descriptor information includes information such as the data types and lengths of the variables, as well as the name of the engine that was used to create the data.

SAS data set
a file whose contents are in one of the native SAS file formats. There are two types of SAS data sets: SAS data files and SAS data views. SAS data files contain data values in addition to descriptor information that is associated with the data. SAS data views contain only the descriptor information plus other information that is required for retrieving data values from other SAS data sets or from files whose contents are in other software vendors' file formats.

SAS data view
a type of SAS data set that retrieves data values from other files. A SAS data view contains only descriptor information such as the data types and lengths of the variables (columns) plus other information that is required for retrieving data values from other SAS data sets or from files that are stored in other software vendors' file formats. Short form: data view.

SAS format
a type of SAS language element that applies a pattern to or executes instructions for a data value to be displayed or written as output. Types of formats correspond to the data's type: numeric, character, date, time, or timestamp. The ability to create user-defined formats is also supported. Examples of SAS formats are BINARY and DATE. Short form: format.

simple index
an index that uses the values of only one variable to locate observations.

SQL
See Structured Query Language.

SQL pass-through facility
the technology that enables SQL query code to be passed to a particular DBMS for processing. Short form: pass-through facility.

Structured Query Language
a standardized, high-level query language that is used in relational database management systems to create and manipulate objects in a database management system. SAS implements SQL through the SQL procedure. Short form: SQL.

union join
a type of join that returns all rows with their respective values from each input table. Columns that do not exist in one table will have null (missing) values for those rows in the result table.

view
a definition of a virtual data set that is named and stored for later use. A view contains no data; it merely describes or defines data that is stored elsewhere.

WHERE clause
the keyword WHERE followed by one or more WHERE expressions.

WHERE expression
defines the criteria for selecting observations.

Index

U

V

W